MONARCH OF THE BUTTERFLIES

Ken Parejko

Acknowledgements

I would like to gratefully acknowledge the
wide-ranging hands-on knowledge and enthusiasm
shared by contributors to the Dplex-l listserv.
I'm especially grateful for the helpful comments offered by
Barb Case, Ina Warren, Drs. Michelle Solensky, Chuck Bomar
and an anonymous reviewer. I remain responsible for any and all errors.

For comments or corrections, please contact:
parejkok@uwstout.edu

Table of Contents

Chapter 1
Life History

"There's nothing constant in the world,
All ebb and flow, and every shape that's born
Bears in its womb the seeds of change."

- Ovid

There are many ways to know the world: through poetry, and art; spiritual inspiration; intuition; logic and reason. And there is science. As a scientist, though I write poetry and enjoy art, use my intuition and try to be reasonable, I am especially fond of what is loosely called "the scientific method." It may not be the most impressive or emotionally resonant way of knowing the world, but in the long run it is the most trustworthy.

As a teacher, I tell my students: Scientists let the world tell them how it works. That's not an easy thing to do, as it turns out. We have our preconceptions, our faulty perceptions, our misconceptions. The road from ignorance to understanding is not a throughway; more like a winding rustic road, full of wrong-turns and detours, some of them offering beautiful vistas, seductive but unreliable. For centuries now we have bent over microscopes, shivered at chilly telescopes, sampled soils on the edges of simmering volcanoes, babied dangerous distillations or wrestled with rebellious DNA gels. As Steven Jay Gould[1] put it: "Most daily activity in science can only be described as tedious and boring, not to mention expensive and frustrating." But the hard work, coupled with the insights of pioneers in every field, funded to a large degree by your tax money, have set at our feet one wonderful gift after another. One by one we can pick up pieces of the world never before seen and with them put together new patterns and new understandings of how it all works. And each new vista brings with it the promise of more to come.

The growth of scientific information has been described as exponential. It is that and more. It was barely a hundred years ago that August Weisman suggested that genes were on chromosomes. Dr. Hugh Iltis related to me and other students in his evolution class at the University of Wisconsin how his father was a friend of Weisman's. We are but two generations from even knowing where the genes are, and now we redesign them. Einstein's three seminal papers were all published just over a century ago. Yet how far we've come since in our understanding of the universe, its beginning and its evolution.

While there are just over 700 species of birds nesting in the U.S. and Canada, there are about 90,000 species of insects already identified, and plenty yet to be named. Some say that insects count for about 70% of all animal species on earth. We know quite a bit about some of them – especially the ones we call pests. But the behavior of most, their ecological interactions with each other, the plants they feed on or pollinate, and the predators which eat them, remain to a large degree behind the veil. Those who feel uncomfortable without the resonance of a mystery lurking over the hill may find that satisfying. It means too there'll never be a shortage of good questions to answer, each a potential career. But it also means we're still on a steep learning curve about the world we inhabit.

In biology the study of an organism's birth, development, and death is called its "life history." It turns out we know quite a lot about the life history of the monarch.

All life histories have something interesting about them. But our monarchs give us something special: these hunky caterpillars who transform themselves into delicate butterflies, ugly ducklings who crawl off to sleep only to wake one morning as lovely swans into a completely new world. Within the hemispheres of our brain resides a terrible knowledge. On the one hand we cling desperately to our self-being but on the other know only too well of our constant becoming. We are who we are, yet we constantly become something else. We cling to the hope that our transformations, so often out of our control, will be for the better. The monarch's throwing off the cloak of the ever-hungry larva, tied by its desires to the mundane world and emerging as a light, free and beautiful butterfly is emblematic of that hope, that our own transformations, the little daily ones and the big ones—birth, romance or marriage, and finally death— will be for the better.

But the process of metamorphosis resonates with us for even deeper reasons than this, which I believe have to do with the very fundament of who we are, who we think we are, and how we perceive the world.

Consider a tree: any tree. The tree you see is not the tree as it is. It is both much more and much less. It is more because we perceive only its surface. We see its bark, the lichens it supports, the filigree of its leaves and in spring or summer its dangling flowers. But we cannot see the practiced orchestra working feverishly away inside it – its enzymes, chloroplasts, intercellular junctions, xylem and phloem, a complexity ultimately beyond our ken. And the veil is even more opaque than that: that tree's DNA, almost certainly like each of ours absolutely and forever unique; its own particular history from seed to sapling, the winds which bent it and rains which watered it, or hailstones which threaded the leaves off it that summer not so long ago; the ancient evolutionary history of its clan, of which it is only a momentary epiphany. Each tree is an infinity we cannot plumb.

But the tree is at the same time less than what we see. Because by knowing of trees, the cottonwoods we've camped under on the high plains, or the western cedars and Douglas firs of the Olympic rainforest; and of wood floors, which welcomed us into the old hardware store of our childhood, or of Birnhamwood moving to Dunsinane which brings to our every meeting of a tree the cold winds shrieking about the thanes of Cawdor; and the squirrels, gray or black or red, chattering and nesting in its crown, which we have encountered in parks in Europe or on the college campus we came to as wide-eyed freshmen; and of the memory of the sweetness of maple syrup as our mother poured it over the steaming Sunday-morning pancakes; with all these, which are uniquely our own and not of the tree itself, we make the tree more than it is.

The poet Wallace Stevens wrote that Marianne Moore had "the faculty of digesting the 'hard yron' of appearance." So with our eyes or ears or nose we ingest the tree, and with our imaginations we digest it. And in digesting it we make it part of ourselves, and in doing so change not only ourselves, but the tree, too, transmuting and transforming it's image while it transforms us.

It's said that though his medium was poetry, Stevens' heart was the heart of a painter. His poems in many ways are about the transformation of the visual into something uniquely and very human. Of painters, I think none is closer to Stevens than Van Gogh. By transforming the everyday world — the peasants sitting down to a meal of potatoes, a pair of old boots, a grain-field with crows – he shows us something we didn't know before about that world, and something new about ourselves, too. Most importantly perhaps Van Gogh's and Steven's imagination reflect to us an image of the vitality at the center of life, which intimacy with nature can also bring us into intercourse with.

Each day for each of us the world offers itself to this substantial transformation. It is the ground we walk on, the stuff with which we make our art, create our gods, and ultimately ourselves by ourselves and in relation to others.

Our word imagination comes from the Latin *imago*, meaning a picture or representation. The Swedish biologist Linneaus applied the term to the adult stage of insects which undergo metamorphosis – so the butterfly is an imago. As we'll later see, certain tissues in the larva are called imaginal discs, because they carry in them a genetic representation of the adult.

Stevens is the great poet of the imagination. With the will of our imagination, he says, we can actively participate in these daily metamorphoses, and in so doing become more than we are; we can achieve a meaning-full relationship with the world. In *The Figure of Youth as a Virile Poet* he says "We say that poetry is metamorphosis..." His poem *Final Soliloquy of the Interior Paramour* is nothing less than a love-ballad to the imagination.

"We say God and the imagination are one.
How high that highest candle lights the dark."

and in Notes *Toward a Supreme Fiction* he says that:

"the freshness of transformation
is the freshness of a world."

Poets are naturally drawn to butterflies. There's a photograph of Walt Whitman holding out his hand, on which a butterfly has alighted – though it's been pointed out that this is most likely a bit of self-promotion by Whitman, the supposed butterfly being no more than a cardboard cut-out.[2] I know of no poet who wrote more of butterflies than Emily Dickinson. She speaks of the butterfly emerging from its chrysalis and putting on its "assumption-gown." Here is her version of the Catholic's Sign of the Cross: "In the name of the bee, and the butterfly, and the breeze, amen." Though a recluse, as though cloistered within her own chrysalis, each day she unfolded the wings of her imagination which became her own "assumption-gown."

Metamorphosis and change are woven into the very fabric of our lives. But necessary as it is, change can be hard. So it seemed apt that in the highlands of Mexico, near the foot of one of the monarch's over-wintering sites, in Valle de Bravo's church of Saint Francis of Assisi – patron saint of nature – I came across the image of *La Senora de Sagrada Grazzio, abogado de las causas dificiles y desparadas*. Our Lady of Divine Succor, intercessor in difficult and desperate causes. As I turned to leave the church I encountered a pair of women, who like so many of Mexico's poor, live rich full lives surrounded by constant and what would be to many of us insurmountable difficulties, on their knees in slow migration toward the altar.

So yes, our little half-gram butterfly not only navigates the treacherous waters of metamorphosis, it then has the gall to attempt a long flight to a mountain enclave it's never visited, there to take on, unprotected from the elements, five months of hard winter. Much more about that, later.

It's no accident that reproduction is fun. Any species which didn't reproduce, no longer is. In the animal kingdom, it's the story of sperm heading off to meet egg, and it's seldom a straight-line narrative. Like us, monarchs start their lives as one cell, the fertilized egg called a zygote. Unlike us, butterflies have hundreds of brothers and sisters, all with the same mother but who knows how many fathers. We will learn more about that – the female monarch's innate polyandry—in another chapter.

An egg is DNA wrapped in a fancy package. Yo' momma's DNA plus yo' poppa's DNA equals.....you! Monarch or not-the-monarch, alike. Like every one of us, because they are sexually reproduced, except for the rare identical twin, every person and every monarch is genetically unique, and in the long history of the universe never has been and never will again inhabit the earth.

Pliny the Elder writes about nearly everything in his encyclopedic *Natural History*. According to him (Book 6, chapter 37), lepidopteran eggs are distilled from the morning dew by the heat of the sun. It is a lovely description, even if not scientifically correct. It does suggest, however, that in his time – the first century C.E. – readers expected not just descriptions but explanations for natural phenomena. Perhaps that is a trait which makes us uniquely human — not just seeing but needing to explain. One wonders if a chickadee carries that burden, wondering where from come the sunflower seeds on the feeder?

Lepidopteran eggs come in many shapes and sizes. Monarch eggs are elongated – prolate — spheres. Technically, they are referred to as oblate spheroids. They stand upright, about the size of the head of a pin and milky-white in color. You'll find them on the underside of milkweed leaves. Under a microscope you can see that the shell of the egg has one or more tiny holes in it called micropyles through which the sperm entered. The embryo inside the egg is alive, and needs oxygen, so the micropyle also serves as an opening through which the embryo breathes. In addition to its DNA, the egg contains yolk, which holds the nutrients for the growing embryo. Because monarchs are poikilotherms, whose body temperatures closely track that of the ambient air, and metabolic reactions are temperature-dependent, (again, much more about this later) development inside the monarch egg goes faster or slower depending on the temperature. So it can take from perhaps four to twenty days for the embryo inside the egg to mature, often four to seven. As the cells within the egg divide and the embryo begins to develop, special homeotic genes control the formation of the embryo. These genes are very similar to those which early on in your own development set down the pattern for you: head, chest, arms and legs.

The development from a zygote to the tiny, first-instar larva which chews its way out of the egg is itself a near-miraculous choreography of genes turning on and off, proteins dancing their way across the stage and out, with a continual chorus of metabolic reactions, all leading towards the finale of this act: the transformation from egg to larva.

According to the philosopher David Hume, "There is no quality in human nature which causes more fatal errors in our conduct, than that which leads us to prefer whatever is present to the distant and remote." Gifted with hindsight and foresight, our challenge is to immerse ourselves in the here and now without getting stuck in it. But the distant and remote, to a tiny first-instar monarch larva, just out of its egg, is of no interest at all. To achieve its destiny, still hidden over the close horizons of the milkweed leaf, it simply hunkers down and eats, eats, eats. It's a simple life, really. No bills to pay, kids to ferry to soccer practice, tests to take or bosses to satisfy. Just eat, eat, eat. And it does that with a Zen-like single-mindedness.

Its DNA directs the larva to eat. It's a job for which its DNA has well-equipped it. Its mandibles are hard and rimmed with teeth-like structures; near the front, shaped more like incisors, and in the back more like molars. With these and the muscles attached to the mandibles, it bites off and chews up chunks of the milkweed leaf. It uses its maxillae and the palps to direct the food towards its mouth. There, special sensory cells tell it what the food smells and tastes like, and if the food's chewed well enough to swallow.

The monarch larva's tubercles – antennae-like structures — are carefully tuned, by natural selection, to pick up and respond to only certain signals. Especially, but not only, the smells given off by a nice young milkweed leaf. Our senses are tuned, too – we only perceive a small part of the visual and auditory spectra which are all around us. Our sense of smell is miniscule compared to that of our best, canine friends. It's hard to say what the olfactory world of a monarch caterpillar is like. Further research into their sensory and neural biology may someday allow us to paint a better picture of it.

Monarch larvae, born as they are on the plant they will be eating, don't need to look for their food. But they do have eyes, with simple lenses, that allow them to form a rough though poorly-resolved image of the world, an image which at least helps them to see the approach of potential predators and if necessary fall off the plant to escape.

Some of you have had the "pleasure" of giving birth. Mothers are more likely to remember their childrens' birth-weights than their own. Perhaps you do know your own birth-weight. The US average is about seven and a half pounds – less than six, if you're a twin. The first-instar monarch larva crawls out of its egg at an average birth-weight of slightly less than half a milligram. That's not much. If you weighed the US average on birth, it would have taken somewhere around 9 million of the newborn monarch larvae to balance the birth-weight scales. Depending again on temperature, and the nutritional quality of the leaves it munches, munches, munches, over the next week or two the larva will end up weighing nearly a gram. Measured as a ratio, by the time it's done being a larva and ready to crawl into its chrysalis, it will weigh almost three thousand times what it weighed at birth.

Okay. Let's do the math. Let's say you were an average infant at birth. That would make you as an adult, well, twenty thousand pounds. Ten tons. Careful where you step! I don't think those stairs, or that floor, will hold you. And better get a truck to haul you to school or work. A big one.

The very first meal an infant monarch enjoys is the shell it crawled out of— a nice high-protein snack. (You can watch a video of this, and many other interesting monarch clips, by searching on *YouTube* for "monarch butterfly.") The tiny larva, barely visible at this stage, often spins a silk pad to attach itself to, so it doesn't fall off the leaf.[3] But the moment it turns its attention to the big, green tabletop it was born on, it becomes an eating machine. Having worked on a dairy farm, and also having raised many monarch larvae, my imagination connected the dots between these chubby chewers munching milkweed and the bigger, four-legged Holsteins I once kept bulked up with hay and silage. Eating and pooping seem to be what both are best at. Cow-poop we call manure. Insect poop is called *frass*, and from measurements done in my ecology classes we determined that for every milligram of weight gain, monarch larvae deposit about two milligrams of frass. Those of you who've raised larvae have likely been astonished, as I was, at how rapidly it can build up.

The monarch larva's food pyramid is not strong on menu choice. To meet its daily requirement of water (for it never drinks) it eats umpteen servings of milkweed leaf. For calories, it needs plenty of milkweed leaf. Nutrients such as proteins, required fats and minerals it gets by eating a few more servings of milkweed leaves. Finally, to ingest nasty-flavored chemicals which protect it from potential predators, it grazes on...milkweed leaves. The term *pharmacophagy* has been applied to animals eating not for basic nutritional needs such as calories or protein, but for the chemicals in the food, needed for various purposes. Most animals select the days' menu not only based on what's available, but what they feel they need. There are reliable reports of deer standing in creeks to eat fish – nature's nutritional supplements. And monarch larvae will sometimes turn up their noses at milkweed. One woman who raises monarchs put a slice of watermelon in with her caterpillars and they quickly left the milkweed and turned to munching its rind – though they may not have been able to properly develop on such a sweet dessert.[4] Final instars also seem to like butternut squash.[5] Experience in raising larvae suggests earliest-instar larvae seem to prefer the particular species of milkweed they first taste, though later instars are more adventurous in their selection of food.[6] Clearly more remains to be learned about what they will and will not accept as food, and the nutritional and growth consequences of these foods.

Mathavan and Muthukrishnan[7] studied a close relative of our monarch, the queen butterfly *Danaus chrysippus*. By reducing how much they were allowed to eat larval development was stretched out from 6 to 18 days. While the pupae produced by the underfed larvae did not take any longer to eclose (hatch out as adults) they were smaller than normal. My ecology students made similar observations, and we also noted the smaller adults these smaller pupae produced.

It's a story we learn as children, how the monarch caterpillar sequesters chemicals found in the bitter milkweed juices to protect it from becoming someone's snack. But that is just the synopsis. Nature has built a complex plot-line, almost poetic in its details, which deserves closer study.

Female monarchs lay their eggs on milkweed plants, whose scientific name is *Asclepias*. Like many plants, milkweed is a chemical factory, producing what are called secondary metabolites to protect themselves from being eaten. We humans make use of these chemicals – nicotine from tobacco, cocaine from the coca plant, caffeine from coffee are only a few examples. Milkweed itself was used by native Americans. The Chippewa (Anishanabwe) of the northern Midwest gave it to lactating women to stimulate lactation – perhaps on the same principle of resemblance which made the mandrake plant magical to medieval Europeans. Roots from milkweed were mixed by the Anishanabwe with roots of boneset and applied as a charm on whistles, to call deer.[8] Other tribes used the root as a laxative and diuretic, and the milky latex for curing warts, moles and ringworm. Mohawks apparently combined it with an extract of jack-in-the- pulpit as a contraceptive – perhaps actually as abortificents, because both contain dangerous toxins.[9]

In food-preference trials, Vickerman and de Boer[10] showed that monarch larvae prefer species of *Asclepias* to other plants. Extracts of milkweed leaves, but not other potential foods, stimulated the larvae to eat, so the leaves must contain some kind of chemicals – as yet unidentified – which act as phagostimulants, the larval equivalent of walking by a bakery.

The toxic glycosides (also called cardenolides) are about forty times more concentrated in the bitter white latex of the milkweed plants than in the leaves themselves. Monarch larvae have been seen, especially in early instars, to cut a partial circle in the leaf they are feeding on, which causes the more toxic and/or sticky latex to drain. Older larvae sometimes cut the petiole which holds the milkweed leaf to the stem, which also decreases toxins.[11] Because the leaf bleeds out its toxins at the cut, while eating the cut leaf the insect is exposed to as little as 10% of the toxins.

But as it so busily eats, the monarch larva is stepping farther and farther out onto thin ice. Depending on the concentration of toxins in the particular plant it munches, it is in fact poisoning itself, and some number of those happy newborns do just that. Not all milkweed species, or even plants of a particular species, have the same concentration of toxins. Zalucki, Brower and Alonso[12] manipulated latex concentrations in milkweed leaves by tweaking them with forceps, which causes the latex to leak and the total concentration of glycosides in the surrounding tissues to go down. They found that larvae that fed on the damaged leaves, with less glycoside, grew about twice as fast

9

as those which fed on undamaged leaves. Some of the larvae became stuck to the latex, and died. Early instar larvae are especially sensitive to the latex. The latex is sticky, and hard to clean from their mouths and heads. Videos showed some, who'd consumed the high-toxin latex on undamaged leaves, becoming cataleptic and falling off the leaves. In similar experiments, first instar larvae grew faster, and experienced lower mortality, when the glycoside concentration of the leaves was reduced by artificially causing the latex to leak from the leaf.[13]

So the larvae, though protected by the glycosides, pay a price — like medieval knights donning heavy armor. By overloading their own detoxifying systems with high-titer milkweed, they can either stunt their growth or in some cases kill themselves. It's not a free lunch!

Individual monarch larvae hatch onto a particular milkweed plant which can have low, medium or high concentrations of cardenolides. One study found the average concentration to be about 50 micrograms of cardenolide per 0.1 gram of common milkweed (*Asclepias syriaca*) dry tissue. Variation among plants however was very wide, from 5 to 229 microgram per 0.1 gram. On average the 158 larvae assayed had a mean of 234 microgram per 0.1 g of larval tissue. In other words on average they concentrated the cardenolides in their bodies by a factor of about 4.7 times. Eastern plants and larvae had higher concentrations than more western. Milkweeds in small patches had more cardenolides than those in larger patches, though the larvae did not. Younger plants, and those in the sun had higher concentrations of cardenolides, as did the larvae found on them. One cardenolide called aspecdioside was found in 99% of the plants and 100% of the larvae. Another, desglucosyrioside, was not found in any of the plants but did occur in 70% of the larvae, who must therefore be capable of synthesizing it themselves, perhaps from other cardenolide precursors.[14] Other studies showed that larvae who eat leaves with high concentrations of cardenolides no longer concentrate it. Instead, they reach a kind of saturation level within their bodies, above which it becomes toxic to them.[15]

That's the thin ice. You have to eat to grow, and like most salads milkweed leaves are low in nitrogen. So you have to eat enough to grow and to meet your nitrogen and other nutritional requirements, but not so much as to poison yourself. Studies show that when the milkweed plants are grown in limited nitrogen, the larvae have to eat more. But you also eat to protect yourself from predators, by sequestering the toxins in your food. Consume too much toxin and you're sick or dead. Consume too little and you either starve, don't get enough nitrogen to grow, or don't have enough toxin to protect yourself should a passing bluebird swoop down to try you out. It's a fine line you walk to get it just right. Some do, some don't. Time again to call on *Our Lady of Difficult and Desperate Causes.*

Of course if every one of the hundred or more eggs a reproductive female laid survived, the world would soon be overrun by monarchs. Most of course do not survive. Zalucki and Brower[16] studied survival of monarch larvae in north-central Florida on the sandhill milkweed (*Asclepias humistrata*). Survival of first- instar individuals in the wild ranged from about 4% to 12%. That's not mortality, that's survivorship! About a third of the larvae were found glued to the leaves by the plant's latex. Some starved because their mandibles were glued together as the latex dried. Survival of eggs collected in the field and hand-raised was much higher. Of 219 eggs collected and hand-raised, 62% hatched into caterpillars, and 64% of eggs hatching into caterpillars were released as adult butterflies.[17] Similar results for survival and longevity have been found for many hand-raised or zoo-kept animals, protected from natural predators and other sources of mortality.

In nature some larvae just fall off the leaf during a storm. Or a newborn larva, just finishing its eggshell snack, might bite down onto a leaf's vein, with its high-titer toxin, and go into a seizure, unable to move or hide from predators.

Evolution naturally has been busy designing predators of the larvae. The monarch larvae monitoring project[18] has shown that various parasitoids, including tachinid fly larvae, also contribute significantly to larval mortality. More on them, and other factors which increase larval mortality, in a later chapter.

For the moment let's focus a molecular microscope on the chemicals the larva is eating and sequestering. They are a mix, a cocktail of sorts. Milkweed's genus name *Asclepias* comes from the Greek god of medicine. Like the Native Americans, Europeans have made use of milkweed chemicals. The term cardenolide comes from their use in treating congestive heart failure, because they increase the contraction of the heart.

These cardiac glycosides are found in a variety of plants, including the foxglove *Digitalis lantana*, from which we get the powerful cardio-stimulant digitalis. Though not found in milkweed, ouabain is a related compound whose toxic effects have been studied as a model of other, less common cardenolides. How is it that monarchs munch cardenolides with relative ease,

11

while other herbivores who might ingest it become so rapidly sick? Modern molecular biology provides us an answer.

Ouabain is toxic to most animals because it inhibits a very important enzyme called ATPase – a protein which helps break down ATP (adenosine triphosphate) and in doing so releasing energy. If we can't break down ATP, we run out of energy – it's like having a laptop with its AC-adaptor but no place to plug it in. Eventually, the cells' batteries run dead.

So if you could evolve an ATPase which was not sensitive to oubain, you could build up higher concentrations of it and stay alive, while at the same time becoming toxic to predators. That's exactly what monarchs have done. Ouabain (and presumably the many related cardiac glycosides in milkweed) binds to the ATPase to make it inactive. But monarchs carry a mutated form of the ATPase. At position 122 of the protein, compared to most lepidopterans, monarchs have an abnormal amino acid – a substitution of the amino acid histidine for an asparagine, which causes the ouabain not to bind to, and inhibit, the ATPase.[19] These same scientists went on to clone the binding-site on the ATPase to which ouabain binds, and put it into human kidney cells. These cells were much less sensitive to ouabain than normal kidney cells. The leaf beetle *Chrysochus*, which also sequesters cardenolides, has the same amino acid change, from asparagine to histidine, at position 122 of its ATPase.[20]

Remember that our larva is busy munching milkweed for its nutrients as well as its glycosides. We don't normally see or think in these terms — because I suppose evolutionarily it would have been too distracting — but in a field scattered with milkweeds, each plant is not only genetically distinct but because it has grown in a local microenvironment, nurture has contributed to its uniqueness. This means that each milkweed has more or less glycosides, and each is overall more or less nutritional to the individual larva, too small to migrate off the plant it was born on to find a better selection at nature's salad bar. Milkweeds can respond to being eaten by dialing up their production of toxins. This ability, called an induced response, is quite common in plants. But depending on its genetics, each plant is more or less able to increase its load of toxins when it finds itself being munched on. If the particular milkweed which is the larva's birthright is low on nitrogen, it must eat more to meet its own protein requirements.[21] But if it eats more, it gets more toxins; so let us hope for the larva's sake that the low-nitrogen plant it's munching is also low on glycosides, and not too swift at dialing them up.

We'll step off the glycoside treadmill in just a moment. One further question though. What happens to the poisons our larva is ingesting? By feeding larvae milkweed leaves, then assaying them for cardiac glycosides, Frick and Wink[22] showed that within two days after ingesting them the toxins wind up in the integument (skin). The midgut tissues and the hemolymph (insect equivalent of blood) on the other hand seem to function merely as transient compartments for them – shopping bags for bringing the groceries home. They also found that a little over 60% of the glycosides were transferred from larvae into the butterflies during metamorphosis, the main sites of storage in the butterflies being the wings and integument.

It may be that not all the poisons a particular monarch larva lugs around with it come from the milkweed plant. Two subspecies of the monarch found in South America are capable of synthesizing some cardioactive toxins even if they've been fed a diet which does not contain them.[23] Evidence mentioned above regarding the glycoside desglucosyrioside suggests North American monarchs make some of their own toxins.

Most species of butterfly larvae, of course, do not feed on toxic milkweed. Bernd Heinrich[24], a wide-ranging biologist who has written much about insects and whose book *The Mind of the Raven* is a revelation of complex corvid behavior, observed that palatable species of caterpillars – those not protected by toxins — tend to always feed on the underside of leaves, nap during the day while feeding at night, and after snacking on a leaf tend to crawl and hide themselves relatively far from the leftovers which might signal a predator of their whereabouts. Less palatable species of caterpillars often don't do this. In other words, palatable species have evolved behaviors which unpalatable species have not had to evolve — or it may be that unpalatable species like the monarch have over time lost those protective behaviors.

Another strategy for survival of palatable larvae is camouflage. Because they are so busy eating out in the open and don't have time or means to defend themselves, many moth and butterfly larvae have evolved creative ways of camouflaging themselves. Some you could easily mistake for pine needles. They even sit that way on the twig, straight-backed as Amish children in a classroom. Others almost exactly resemble a wet glob of bird poop. Yet others look like broken twigs hanging off a branch. Sometimes instead of camouflage they are masters of disguise. The head of one (*Hemeroplanes*) is a perfect image of a minute snake. Another looks like a snail. When threatened, *Cerula vinula* turns its equally comical and frightening red face at you. Many have sharp and sometimes toxic spines.

But our *Danaus* larva isn't camouflaged, is spineless and doesn't resemble anything other than itself, a very brightly-painted fat and tasty caterpillar. It uses its toxins to protect itself; and it uses its bright colors to advertise that it is poisonous. Its broad, bright bands of yellow, white and black are called by biologists aposematic coloration. Such bright colors are also worn by poisonous snakes like the coral snake, by the gila monster, and by many poisonous spiders with bright yellow or red markings. Aposematic coloration simply says: Stay away!

The adult monarch's toxic qualities have led other species, such as the adult viceroy butterfly, to mimic it. You'd think the same strategy would work for other species of larvae, too, but there are only a few known examples of larval mimics of unpalatable larvae. The black swallowtail larva mimics the monarch larva, but is itself unpalatable. The clouded crimson mimics both. The pipevine swallowtail, which like the monarch sequesters toxins, appears to be mimicked by the great spangled fritillary. Female pipevines appear to be born with more toxins than males. Over their lifetime they seem to lose these chemical defenses, perhaps by depositing some in the eggs they lay. Older female pipevines, low in chemical defenses, are in a sense mimics of more poisonous adult males. [25]

Not all monarch larvae are identically colored; again, nurture plays its tune on DNA's keyboard. Monarch larvae raised at cooler temperatures have more black and less white and yellow pigment than those raised at warmer temperatures, probably to increase solar heat gain on chilly mornings.[26] In a similar way in some species of butterflies adults born in cooler springtime are darker than those born during high summer.

We'll look more closely at the complex co-evolutionary interplay between the monarch and the milkweed in the *Ecology* chapter of this book.

Time now, though, to pay the piper. Dr. Miriam Rothschild was one of the first biologists to study the relationship between the monarch's sequestered chemicals and the aversion predators have to them. Her uncle Walter Rothschild possessed what was likely the world's largest collection of butterflies and moths, some two million specimens. Miriam was a prolific letter-writer, composing a letter every day for 35 years to her cousin, head of the Rothschild bank in Europe – to him alone, more than 13,000 letters. During World War Two the Rothschild estate she lived at housed American airmen, including for a time the already-famous Clark Gable.

Miriam Rothschild had little formal training in biology. She believed in studying nature, not books. She once said "The types of tests devised by the appropriate authorities in Britain today assess the size of the child's bottom, rather than that of its head." She became an expert on the reproduction and taxonomy of fleas, having mounted at least 60,000 of them on microscope slides. Unable to breed them in a laboratory, after discovering that the female flea's reproductive cycle was synchronized with that of the rabbit doe, she was able to breed them like…rabbits.

She is considered by many to be the founder of the study of chemical ecology. Her work on pyrazines, often found in toxic prey such as our monarch larvae, showed that the unique odor of pyrazines actually helps improve an animal's memories – a good thing to do if you want to impress someone who's just gotten sick to not eat you or your kind again.

Dame Rothschild published more than 275 scientific papers, and was awarded honorary doctorates from Oxford and Cambridge. She was well into her sixties when an eye disease prevented her from continuing her work on fleas. So she returned to her earlier interest in butterflies, which she described as "Dream flowers, childhood dreams which have broken loose from their stalks and escaped into the sunshine. Air and angels!"

In her eighties she campaigned with Prince Charles for replanting England's wild places into wildflower havens for butterflies, and under her influence Highgrove, one of the Prince's properties, became one of the first to feature not expanses of lawn along its drives but swaths of wildflowers. When asked among all her life's accomplishments what gave her most joy in her life, she answered having a family.

This is from one of her letters, written on Christmas Day, 1989, when she was 82:

"I am playing about with wildflower promotion, a bit of propaganda on television, a book trying to combine poetry, painting and natural history, a film with David Attenborough, a few vague experiments with quails and pyrazines, but really doddering along, falling over and rather surprised and glad to still be around. I can occasionally look down the microscope but pay heavily for it afterwards with a smashing headache. What I've been trying to look at is the arrangement of sperm within the spermatheca of fleas. This knowledge will have no effect on man's destiny but it intrigues me personally. I am confident no one else alive today knows how sperm are arranged in a flea spermatheca - or cares. And this idea appeals to my perverted sense of discovery."

We scientists stand on the shoulders of giants, and Dame Miriam Rothschild, who died in January 2005, at the age of 96, is one of them. There's a wonderful interview with her, David Attenborough and Prince Charles at http://www.abc.net.au/rn/science/ss/stories/s1331040.htm

The first instar monarch to crawl out of the egg is very, very tiny. Almost microscopic. Because it lives inside an exoskeleton, it grows in increments rather than smoothly. To grow, it must like a snake first shed its skin. It does this four times before becoming the final or fifth-stage larva (caterpillar.)

Myron Zalucki[27] reminds us that "first instar caterpillars are not simply small versions of later instars...it should not be assumed that because you know the biology of a fifth instar, you know the biology of the first instar." Well, no surprise there. A fifth-grader is not simply a bigger first-grader either.

The old hard head-capsule of a larva detaches itself from the rest of the cuticle (skin) just before the molting process is about to start. That's one of the first signs of an impending molt, should you want to watch the process. The new, yellowish head-capsule forms on top of it. Sometimes as the larva completes its molt you can see it shaking its head back and forth, or with anything handy rubbing itself free of the old head-capsule.

Each time they molt, the larvae shed their skin and grow. After their last molt, into the fifth instar, it's time to prepare to become a pupa, the next stage of their life. That means storing up enough food, of the right kind, and water, to become a butterfly. As the butterfly emerges from the chrysalis, it pumps its wings up with water it's stored in its abdomen, so towards the end of its larval life-stage it's eating as juicy of milkweed leaves as it can.

Hormonal changes inside the mature monarch larva also speed up. It's time for the special tissues called imaginal discs to step up to the conductor's platform and direct the orchestra. These imaginal discs began to take shape in earlier larval stages, and with proper hormonal cues now become the scaffolding for the major anatomy of the adult butterfly – one disc for each leg, one for each of the four wings, etc. First discovered in fruit-flies by Ernst Hadorn, a Swiss developmental biologist, the discs play a pivotal role in metamorphosis. Like a teenager new hormones flood the larva – growth hormone, e.g. – while others like juvenile hormone fade in their final decrescendo. These changes take several days, but if you've raised monarch larvae, you can recognize when they are near the finale and ready to pupate – they develop an obsessive wanderlust, walking in search of the right place to hang themselves.

Yes, to move onto another plane of existence, they hang themselves. Having found just the right spot – criteria as yet unknown to us — they attach themselves with a drop of homemade super-glue attached to their butt-end. They hang for a while in the shape of the letter "J." Watching the next hours in their life, as special enzymes dissolve their skin and build around them instead the chrysalis, is as unforgettable as watching a birth.

Pliny, who died in the eruption of Mt. Vesuvius on August 25, 79 A.D., describes the life history of a butterfly (*Natural History*, Book 6, chapter 37.) First the egg is formed from the sun's action on a drop of dew. "From this a small grub afterwards arises," he writes, "which, at the end of three days, becomes transformed into a caterpillar. For several successive days it still increases in size, but remains motionless, and covered with a hard husk. It moves only when touched, and is covered with a web like that of a spider. In this state it is called a chrysalis, but after the husk is broken, it flies forth in the shape of a butterfly." Except for the bit about the sun acting on dew, quite accurate, actually.

For a 21st-century record of the pupation events, see the nice set of photos at http://monarch.org.nz/monarch/2007/10/25/larva-to-pupa-to-butterfly.

In November, 2009 three monarch larvae were launched into space on the space shuttle Atlantis. (see www.monarchwatch.org/space).The purpose of the experiment was to learn the effect of development in essentially a no-gravity environment. Here's what was learned. Without gravity, larvae behaved quite normally, clinging to their food, eating, and if they did anything different, it was to spend less time eating at the top of the food provided than did the control larvae on earth. Their "J" shape, without gravity to stretch their body out, was more of a curly "C." Because of problems attaching themselves to a substrate, all three in space became "floating pupae." When emerging from the pupae (see below) the butterflies were not able to expand their wings in a normal way, though two of the three were able to fly, within the confines of the enclosed area provided them. There was some discussion online whether the results of the experiment merited sending three larvae into space, knowing that after transforming they would be unable to feed or reproduce.

We come back from space now and return to our pupa inside its chrysalis. The chrysalis of different kinds of butterflies are quite different, and can often be identified by species. In fact, for those of you who raise or find a monarch chrysalis, it can even be sexed![28] Because the pupa inside is very edible and unable to protect itself, most moth cocoons and butterfly chrysalises are painted in camouflage colors. The monarch chrysalis is a lovely pastel green. But it's design holds to me one of life's great mysteries – three tiny dots of gold spread across its top, and a line of golden droplets like an expensive necklace below. Hard materialists say natural selection is chance alone, without a guiding hand. Some find in the aesthetics of the monarch chrysalis an argument for intelligent design. No good biological explanation for their existence has as yet come forth.

For the next week or so, though from the outside the chrysalis is all an image of meditative quietude, inside it is undergoing a massive remodeling project. Though most of the larval cells themselves actually survive this rearrangement, some undergo a process called programmed cell death. This is the same process that happens to a tadpole's tail as it turns into a frog, and the tissue between our fingers at about fifty days of our own development, when our hands turn from resembling ping-pong paddles into five-fingered organs. Organelles called lysosomes release digestive enzymes which digest the larva's cells from the inside out. The molecules of life – proteins, fats, carbohydrates, nucleic acids – released by the kamikaze cells are sopped up by the surrounding tissues and the cells which have remained alive. Under the direction of the imaginal discs, which direct a complex composition in the medium of DNA expression, first the pupa forms from the larva, then from it the butterfly itself appears.

This transformation is a fascinating, complex and dangerous process. A lot can go wrong. That it is so often successful is quite remarkable.

Just as nations, schools and other institutions, and we as individuals must make decisions about how to use limited resources, so do transforming butterflies. One question that arose in some biologists' minds was: is there a competition within the chrysalis for these resources? The wings of a butterfly are its largest organ, making up some 20% of its body mass, a heavy demand on resources. In one experimental study the researchers surgically removed from late instar caterpillars the pair of imaginal discs which direct formation of the hind wings of the Buckeye butterfly *Precis coenia*. The fore-wings, thorax and legs which grew from these caterpillars were significantly larger than those which hadn't been manipulated. This suggested that there was competition within the pupa for limited resources, and if one part got less, the others got more.[29] These results also remind us that when a gene undergoes a mutation which could potentially be adaptive, if that adaptation steals resources from another part of the individual, the change might no longer be adaptive. Every monarch, whether caterpillar, chrysalis, or butterfly, is the result of millions of years of evolutionary tinkering. And like a finely-tuned sports-car engine, changing that tuning is more likely to worsen matters than improve them.

If the imaginal discs which the researchers removed are placed in a petri dish with nutrients, they fail to grow. But when the steroid hormone 20-hydroxyecdysone (the hormone which also controls molting) plus the insulin-like neurohormone bombyxin, produced by the pupa's brain, were added to the nutrients, the imaginal disk continued to grow.[30] No real surprise there: inside the chrysalis, the transforming butterfly is awash in hormones, and if it dreams at all it seems to me they must be sea-dreams.

A fair amount of work has been done on the ebb and flow of gene expression in the late-instar larva as it prepares to pupate. There's a homeotic gene called distal-less which plays a big role in the development of the particular wing pattern each species shows.[31] This gene is one of the most important in limb formation and has been well-studied in the fruit fly and other insects. Analogues of it play a part in limb formation in vertebrates like us.

One study looked at the heart of the pupa entering metamorphosis, and found that it indeed survives the process, though changed.[32] Another paper reports monitoring the heart-beat of the developing pupa, at first irregular but uninterrupted when nearing emergence into the butterfly.[33]

The next step in the life of our monarch is as fascinating as the transformation of the larva into the chrysalis. Perhaps you've been fortunate enough to have watched the chrysalis turn transparent, and see the nascent butterfly all folded up inside. That's a signal that it is about to be born – eclose, in biological terminology. And the eclosion of an adult from the chrysalis is, like all births, a marvelous thing to see, a visual choir of glorious hope and redemption.

It takes only a few hours: by scraping away at its package, the adult breaks free of the chrysalis.

The first morning of a new world.

The Roman poet Ovid recorded his countrymen's fascination with stories of metamorphosis. Birth and rebirth are themes common to much of the world's art and literature. The monarch caterpillar rests like a phoenix in its tomb, then breaks out into the light of day as a new, almost spiritual, creature. In one ode to the phoenix (*De ave phoenice*, attributed to the Roman poet Lactantius) the connection between phoenix and butterfly is made overtly.

After the fire which consumes her, the phoenix becomes a kind of legless larva:

> "...gathering into what looks like
> a rounded egg, in which she is remade
> into her former shape, bursting forth
> from her shell and springing to life as the phoenix;
> as does the larva, in the countryside,
> fastened by threads to a stone,
> to become a butterfly."

Some have seen in this transformation Christ's breaking out of the tomb. Farther back into Egyptian times the scarab beetle was given a special place of honor. The scarab larvae dig long vertical shafts into the earth where in underground chambers they transform themselves into very mummy-like pupae. These shafts and chambers are very similar to the mastaba tombs of Egypt's Valley of the Kings. Mummies were Egyptian – and Incan — pupae, supplied with food, little ships and maps to guide them to their own rebirth. Butterflies – some identified as the North American monarch's cousin *Danaus chrysippus* – are found on the walls of Egyptian tombs going back more than 5000 years.

But the chrysalis is not a tomb, it is a dressing-room in which the larva changes into its assumption-gown. The door flies open, and where was a chubby stomach on stubby legs now instead is a dream-flower, all air and angel. But as it leaves its temporary tomb, our fresh-minted monarch isn't quite ready for its first flight. Because the pupa is like a mummy, the dream-flower it holds is tightly wrapped. Before it flies it must unfold its wings. This it does by pumping water, stored in its abdomen, out through the wings' fine network of veins. And once spread, the wings must be given time to dry. Interrupting this process can derail it — the new butterfly might never fly. Again, it is walking a fine line of vulnerability.

Do enough neurons and interneural connections survive the storm of metamorphosis to allow memories of its former life to flicker like ghosts around the edges of the new butterfly's mind? We don't know, and perhaps we never can. As a biologist I'd guess only those memories would survive which are in some way evolutionarily advantageous. One study shines some light on this question. Female moths and butterflies of many species most often lay their eggs on the species of plants they ate while larvae. That's as it should be – ovipositing on just any plant might be the waste of an egg. How does the adult know what plant to lay its egg on? It could be completely hard-wired. Or does the memory of what it ate as a larva survive metamorphosis and somehow affect its choices? Larvae of the diamondback moth were fed one of three different kinds of food. After eclosing, females showed no preference for ovipositing on the particular food they'd been raised on. Nor did adult moths derived from larvae raised in the presence of neem, a chemical found in the neem tree and an ovipositing deterrent, react any differently to that chemical than from those raised without it. These results suggest that oviposition preferences are made not based on memory but are hard-wired into the adult.[34]

However some memories do seem to survive the extreme make-over of metamorphosis. Martha Weiss, at Georgetown University, provided tobacco hornworm larvae with a strong-smelling chemical (ethyl acetate), then shocked them. They soon learned an aversion to the ethyl acetate, an aversion which a large majority of individual hornworms maintained right through metamorphosis.[35]

The new-minted butterfly wakes from its sleep into the first dawn of a new world, which looks different, tastes different, smells different, and feels different than the world it once knew. It has, according to the words of Colossians 1:10-11, "taken off the old self with its practices, and...put on the new self." It has been given new eyes, a new nose, new almost everything. The world itself hasn't changed, but with their new eyes every summer all across the earth uncountable millions of new worlds are first seen, tasted, felt and smelled.

Perhaps even before it ecloses our butterfly already has had a glimpse of its new world, as though through a glass darkly. Hours before it breaks free, the chrysalis has already become transparent. Perhaps that is why it becomes transparent — to give it a glance of the world it is about to enter. If only we could get inside the brain of a monarch about to eclose, and marvel with it as that new world unfolds.

The butterfly's sensory toolbox is a different toolbox than the caterpillar's. Evolution has given each the tools most adapted to its way of life. A taste for bitter milkweed leaves becomes instead a taste for sweet nectar. Lets open up the butterflies sensory toolbox and see what's inside.

Moths mostly fly at night, butterflies in the day. One kind of butterfly that flies at night are the hyelids, which may be the "missing link" between the evolutionarily older moths and the more recent butterflies. The ability to hear ultrasounds – high frequency chirps made by bats – is common in moths. This allows them to know when a predatory bat is nearby, and take evasive action. Moths seem to have developed ultrasonic hearing at about the same time as bats developed echolocation (about 60 million years ago.) Hyelids, nocturnal butterflies, have been found to have ultrasonic ears on their wings, which they use to escape bats. These tiny ears have an eardrum set in a narrow passageway that resembles a rabbit's ear. It's been suggested that heavy predation by bats may in fact have forced some of them to fly during the day, some 50 million years ago, and in time these diurnal moths evolved into the butterflies. A certain organ of unknown function called Vogel's organ, found in various of the Papilionid butterflies, may be a vestigial remnant of their ancestor's ultrasonic ear.[36] Monarchs are diurnal. There is no evidence they can hear in the ultrasound, though there is some evidence butterflies in general have a sense of hearing, which has been very little studied.

Not only hearing, but the visual abilities of animals are closely linked to their way of life – dogs see different colors than we do, as do the birds and the bees. If you're a flying creature, distance vision is important; it's almost certain the butterfly is less myopic than the caterpillar. In general monarch vision appears to be quite poor. Chip Taylor (Dplex-l list – to sign up, go to http://www.monarchwatch.org/dplex/index.htm) reports watching a number of them as they stream down from their hibernation site fly straight into a shrine. Others have reported watching monarchs fly straight towards an object, then at about a meter's distance dodging it. We need to know more about the visual acuity of the monarch at all stages. We'll look in some detail, however, at the monarchs' visual skills which help it to migrate, in a later chapter.

In addition to its senses, evolution has issued the eclosing monarch the body it needs to survive. Wings for flying, legs for crawling (and tasting), a new digestive system to digest not leaves but nectar, the right set of senses, a long proboscis for reaching nectar, reproductive organs.

So it's no longer acceptable to just eat the table on which you walk. No, now you have to find food. Find flowers with nectar. Note to self: Look for flowers.

Well, the new toolbox helps. The world is a colorful place, and flowers make it more colorful. They're full-color ads, really. Targeting insects mostly, inviting them to stop in for a meal – and while you're there would you pick up a packet of pollen for me?

We'll see in another chapter that butterflies, like many insects, can see light's plane of polarization. Their eyes have been designed to do that. What evidence is there they can see different colors? Foraging Japanese swallowtail butterflies were trained to feed on sugar solutions placed on differently-colored disks. They very quickly learned which color the nectar was in, demonstrating both good color vision and an aptitude for learning.[37] The pipevine swallowtail shows similar abilities.[38]

Butterflies have actually shown remarkable ability to learn, given the size of their brains. Female pipevines are able to keep more than one thing in mind, something I find harder and harder to do. When trained to oviposit on a particular color of leaf, and at the same time trained to find nectar in a different color of flower, they regularly – though not always — landed on the correct color according to whether they meant to oviposit or to drink. [39]

Maureen Stanton[40] studied short-term learning in three species of butterflies belonging to the genus *Colias*. These females lay their eggs on legumes, though especially when they are new to ovipositing, they'll mistakenly land on plants that aren't legumes. During a long day of foraging for oviposition plants they get better at recognizing them and make fewer mistakes as the day goes by. But they also need to catch a drink of nectar now and then, and after searching for flowers (mostly non-legumes) to feed on before going back to egg-laying, they once again make more mistakes. Sound familiar?

The advantages of this kind of learning in insects were shown by a study which presented grasshoppers with nutritious or non-nutritious food, either placed in a predictable environment, where they could learn to go to the nutritious food source, or in a random environment where learning was impossible. Individuals grew faster in the predictable environment, which allowed them to learn and select nutritious food.[41]

Though we can't, many insects can see in the ultraviolet. If you put on ultraviolet-sensitive glasses, the world of flowers leaps into a whole new dimension of beauty. There are lines and circles on flower petals written in ultraviolet ink invisible to us but which direct pollinators inward towards the nectar and pollen. One species of lupine, *Lupinus pilosus*, is pollinated by bumblebees. After a bee visits a particular flower, the flower changes color, indicating to other foraging bees that the nectaries, like an ATM out of cash, have been emptied. Manually manipulating the pollen-release mechanism of the flower causes the same color-change, and bees preferred flowers which had not yet changed colors.[42]

Flowers are butterfly restaurants. What criteria does a butterfly restaurant critic use for rating these restaurants? It must have nectar, preferably 20-25% sugar, within reach of the uncoiled proboscis. But the nectar should have a dash of amino acids, too, and some nutritious minerals. And a comfortable landing pad to rest on while eating would be nice.

Though nectar is a favorite food, butterflies will eat other things to meet their nutritional requirements. They've been seen drinking tree sap, sopping liquid from wet soil, gourmandizing on pollen, rotting fruit, mushrooms, carrion, dung, bird droppings, urine, sweat, and slug slime. And that's only page one of the menu. There are others, some positively unappetizing (to us.) They can't chew their food but if they can drink it or lap it up in some way, and it's got what they need, they're not shy at the table.[43]

Males meanwhile also look for the chemicals — or their precursors — they'll need to either attract a female, or build their nuptial gift (more on that, later.) So it's most often males found in "puddle clubs," clustered at puddles of urine, feces, or carrion. Males of a cousin of the monarch, the queen butterfly, land on the dry head of a floss-flower, moisten it with saliva, then sop up the mix of chemicals they need to synthesize their sex pheromones. That's because if they don't got enough sex pheromones, they don't get enough....well, you get the idea.

When we first learn about animals most of us are told there are two kinds: warm-blooded and cold-blooded. The blood from one feels warm, from the other cold. It's true, on a chilly May morning the blood of a rabbit is warmer than the blood of a snake. But later that afternoon, if the snake's been basking in the sun, its blood might be warmer than the rabbit's.

So let's improve our definition of warm-blooded. Warm-blooded creatures have body temperatures which don't vary, while cold-blooded body temperatures track the ambient (environmental) temperature. Better, but still not right. We're warm-blooded, you and I, and our own body temperatures do vary, often by several degrees Fahrenheit over the course of a day. Most of us are coolest in the morning, after our metabolism's been dialed down while we slept, and warmest in the afternoon, after we've been busy all day. And of course if we have a fever our body temperature can go up pretty drastically.

Meanwhile many "cold-blooded" creatures can have body temperatures that are quite different from ambient. If it's really hot in the desert, say one hundred forty on the sand, insects, snakes and lizards find a shady place to cool themselves, or burrow into cooler earth. Or if it's a cool desert morning, forty degrees Fahrenheit, they might bask in the early morning sun, picking up energy from its rays and warming themselves. They have a repertoire of strategies for warming, which includes physical activity that generates metabolic heat. Many of these "cold-blooded" creatures have anatomical or physiological means of using or losing metabolic heat. Moths studied by Bernd Heinrich can shunt blood flow to dump metabolic heat if they need to, or keep it near their wing-muscles, if they need it.

So, like most things in biology and life in general, there are exceptions to prove the rule. The rule, then, restated, becomes this: Warm-blooded creatures (endotherms, also called homeotherms) have body temperatures which vary less, usually considerably less, than cold-blooded creatures (also called ectotherms, or poikilotherms.)

Insects like bees and butterflies can't fly when their body temperatures are too cold — their muscles and neurons just don't work well. So as the thermometer follows the setting sun bumblebees sometimes get stuck on, and overnight on, the last flower of the day. But there's plenty of evidence that insects shiver, flutter their wings, or bask in the sun to warm up on chilly mornings.

If they could set the outdoor thermostat, what range of temperatures would a monarch prefer? Well, as larvae, they'd want it somewhere between 10C and 35C (about 45 – 95 F). Outside those temperatures they can't survive for long. Adults can survive a much greater range, from as low as -14C to 41C (7 to 106 F.) But they're only able to fly between about 12 and 40C (54 and 104 F), and they really don't like temperatures above about 33C (90F.) In spring, once the temperature in the more southern states they're passing through rises consistently above that, they head north. On really hot days they run the risk of overheating, and will spend considerable time resting in the shade. If they absolutely must fly in those temperatures, they prefer to glide rather than flap.

Females lay their eggs on the bottom side of milkweed leaves to be less visible to predators, but they choose plants that are exposed to the sun's warmth for a good part of the day, so the eggs develop more rapidly. Larvae appear to bask, and can gain eight degrees C or more doing so, reducing their development time by as much as half. If they become overheated they've been seen to crawl down the plant onto the cool earth, or into the shade. The larvae have also been seen to behave as though they are sleeping. They'll sometimes hang their heads and go into a quiescent period. Barbara Case (Dplex-l list) has posted a photograph of several of the caterpillars in a shoebox nicely lined up, as evening comes on, dozing off like tired monks in a monastery at: http://www.flickr.com/photos/8812810@N02/773429215

When they're flying, monarchs generate significant quantities of metabolic heat, just as we do when we're exercising. A butterfly's excess heat is efficiently carried away by air currents as it flies through them, so unless its very warm (above 90 F), they don't overheat. On the other hand there are times when they do need to heat themselves up to operating temperature. Like their larvae, many butterflies bask in the warm sun to heat themselves. I recently watched a mourning cloak do just that, on a mid-March sunny day in northern Wisconsin. It had found a slope of lawn that set it at just the right angle to optimize its solar gain.

Adult monarchs also bask. While doing so it appears that their wings do not act as solar panels; basking is warming their body, not their wings, though the wings do help produce still air around them to prevent convective heat loss. The wings are also used as reflective panels to reflect solar energy to the body, where there is a coat of heavy hair-like scales on the thorax to act as insulation and retain the heat. At the same temperature, males seem to have higher metabolic rates than females, and reproductive individuals higher than diapausing (migrating) individuals (Dr. Chip Taylor, Dplex-l list.)

Moths, mostly nocturnal creatures, find little gain from basking in moonlight, and have instead evolved the ability to shiver to warm up. Monarchs are one of the few butterfly species known to shiver. Shivering warms the body about the same rate and uses about the same amount of energy as flying.[44] At around 15 C (60 F), by shivering they can warm their thoraxes by about two degrees Fahrenheit a minute.[45] Once their thoracic muscles are properly warmed they can launch into the air, confident that flying itself will keep them warm.

Because development is so temperature-dependent it's possible to make some interesting predictions about how long development will take and how many generations can develop at different latitudes. A degree day is one day one degree above the lowest development threshold for an insect, which for monarchs is 11.5 degrees C (53 degrees Fahrenheit.) Since the egg requires about 45 degree days to hatch, at 12.5 C it would require 45 days – a long time! You can do the calculations for hatching time at different temperatures, and if you're a teacher, it's a good way to link math to biology. Though there seem to be more than one opinion on the exact answer, it takes about 200 degree days from egg to pupation, and 700 degree days for a newly-laid monarch egg to hatch, develop as a larva, pupate, eclose, and for that female to lay her first egg. Seven hundred degree days, egg to egg. In Winnipeg, which for the years 2003-2006 averaged around 1500 degrees days up to the end of September, that's a total of two generations. In St. Paul it's perhaps three, in Dallas more like six. An unusually cold summer can change all that, and global warming, should it continue, is likely to also.[46]

If our new adult monarch – first morning of a new world – is in the first generation of that summer's to eclose in the northern part of the U.S., it will become a reproductive adult. Its children are likely to, also. But as the long days of high summer wane towards August and September, it would do no good to try to reproduce. Eggs laid that late would not become adults before the hard killing frosts of October. Instead of reproducing, generations which eclose in the later parts of summer migrate. And migrating monarchs live much longer than reproducing individuals – reproductives, two to six weeks, migrators six to nine months.

How they "decide" which to do is an interesting story.

It's a story about hormones, mostly.

Juvenile hormone (JH), for one. In us, what we call the "sex" hormones – estrogen and testosterone, e.g. – play a big part in puberty and achieving reproductive maturity. In insects it's juvenile hormone that plays that role; without it, the gonads and auxiliary organs simply don't mature. Female monarch ovaries grow rapidly after eclosion, for about a week, as do the male sex organs. This is correlated with an increase in JH, which peaks in females three days after eclosing, then gradually declines.[47] In females the brain plays a part in sexual maturation, by producing hormones which stimulate the production of an important yolk-protein, vitellogenin.[48]

But maintaining high amounts of JH, ironically, accelerates old age, presumably by increasing metabolic rate. The increased lifespan of migratory monarchs appears to be due to down-regulation of juvenile hormone. Injecting JH into migratory individuals decreases their lifespan, and preventing JH synthesis by removing the corpora allata of non-migratory individuals increases their lifespan by up to 100 %.[49]

Monarchs who eclose in the shorter days of early August, when the milkweed quality has begun to decline and the nights are cooler than they had been, produce less JH. Less JH prevents them from becoming reproductively mature, and kicks them instead into migratory behavior and the longer lifespan that comes with that.

In Minnesota many individuals become the migratory type by the third week of August. By early September all or nearly all of both sexes are in reproductive diapause.[50] Monarchs fed old milkweed reached diapause before individuals fed young plants. There's a kind of hurry-up, we need to leave quality to this switch. Even development from egg to adult speeds up near the end of summer, probably because of lower titers of JH.[51]

Two additional hormones that move a monarch to be migratory are neuropeptide hormones, small proteins produced in the brain. One, adipokinetic hormone, elevates fats in the hemolymph, useful for fueling migration. The other is monarch diuretic hormone, which accelerates water loss after eclosion. To the migratory individual, any excess water left after the wings have been expanded is just extra weight to carry on the long trip ahead, weight which is not as critical if you're reproductive.

The switch from being a reproductive individual to instead becoming migratory, controlled by hormones, seems to depend on a number of environmental clues. The decreasing day-length of late summer seems most important, with the fluctuating temperatures which occur at high latitudes at that time and decreasing food quality also playing a part in urging diapause.[40]

In Christian theology, death is more than an unending bodily diapause. It is a kind of transformation, a kind of undressing, as we take off the lineaments of mortality to put on instead our "assumption gowns." But death also reminds us that not all transformations are for the better. Peace sometimes becomes war; compassion, hatred. Death came every day, more common than the dawn, to Nazi concentration camps. The human mind and spirit can wither before the body. The world held little beauty to the inmates of the camp called Terezin, who experienced instead disease, death, and a weary day-by-day grinding down of body, mind and spirit. Some 15,000 children suffered and died there, or after being shipped to a death-camp. Only a hundred survived, none younger than fourteen. While at Terezin some communicated their feelings in paintings or sketches on walls or scraps of paper. Friedl Dicker-Brandeis, a small, fragile woman, dedicated her life to teaching the children of Terezin to find what little joy they could through art. Others wrote their feelings in short but touching poems. Pavel Friedman[52] was twenty-one when on June 4, 1942 he wrote this poem. He would live a little over two more years yet before dying at Auschwitz.

The Butterfly

The last, the very last,
So richly, brightly, dazzlingly yellow
Perhaps if the sun's tears would sing
against a white stone...
Such, such a yellow
Is carried lightly way up high.
It went away I'm sure because it wished
to kiss the world good-bye.

For seven weeks I've lived in here,
Penned up inside this ghetto.
But I have found what I love here.
The dandelions call to me
And the white chestnut branches in the court.
Only I never saw another butterfly.

That butterfly was the last one.
Butterflies don't live in here,
in the ghetto.

In late 2009, the Holocaust Museum announced it was collecting 1.5 million handmade butterflies to represent the same number of children who died in the Holocaust.

Butterflies are dream-flowers, emblematic of beauty and freedom and unending possibilities. The choices made for monarchs – to undergo metamorphosis, to become reproductive or to migrate to a distant land — and the choices they themselves make – which plant to oviposit on, who to mate with, where to spend the night or which flower to sip from – remind us of the opportunities life gives us to transform and fly. As a species, it's time we thought about who we are and who we've been and who we want to be. Perhaps its time to choose which we'd rather be, the single-minded eating machine or the orange and black dream-flower, transform ourselves accordingly, and in so doing free other species from the death-camps we sometimes transport them to.

29

Chapter 2
Reproduction

"Two butterflies went out at noon
And waltzed above a stream..."
- Emily Dickinson

Friday night in a college town. Young men and women streaming to the bars. In my town there's one doesn't mince words: The *Meet Market*, it's called. They walk in one by one only to emerge later – some at least – two by two. Familiar behavior to biologists: the bars are courtship leks, like the arenas where prairie chickens, capercaillies, midges or some butterflies gather to dance and display and choose who is impressive enough to carry my genes into the next generation. In butterflies, it's called hill-topping, because that's where it usually occurs.

I drove a quiet rural road one lovely summer's evening in northern Wisconsin, where a fallow hillside blossomed with wildflowers while behind it the sun set in an orange blaze. Dozens of butterflies streamed across the road towards the field where, like so many flags flying on a field of battle, orange and black butterflies arranged themselves high up on tall stalks of grass.

As I pulled over to watch, a hundred or more butterflies like so many courtiers danced their dance of courtship, rising and falling in the wind, finding a lover and off to the sidelines to mate, or dancing out of play down the wind. The role pheromones released by the male or the female play in monarch courtship remains a question of some dispute among monarch biologists. Perhaps the butterflies I came across had congregated in this particular field because it was a good nectar source. But having gathered there, like a chattering crowd in a busy restaurant, their attentions seemed as much on one another as on the food before them.

I wandered out among them. They rose and settled around me, too busy with the business at hand to notice me or care. I was amazed at how close I could come to touching them before they released from the stalk of grass and let the wind blow them away.

In the golden light of the setting sun it was a remarkable sight, and like watching thousands of sandhill cranes rise off the Platte in a south Nebraska sunrise, or the courtship acrobatics of a pair of eagles high in the air, one of those unforgettable gifts of nature.

Hill-topping in other species of butterflies resembles human *Meet Market* behavior in other ways. Rival males sometimes face off in a slow ascending flight by which they assess each other's strengths and weaknesses. Or they perch on leaves and, like fragile sumo wrestlers, engage in pushing bouts until one is dislodged and forced to fly off.

They say Benjamin Franklin was truly taken by the court of Louis XVI, and the court by him. I'd love to slide back to 1776 and join him in that heady give-and-take. For me the more plebian *Meet Market* rituals have lost their attraction. But I'd sign on the dotted line, or stand in queue all afternoon, to once again join those courting monarchs.

Reproductive behavior is crucial to a species' survival. Birds do it, the bees do it, even monarchs and Dame Rothchild's fleas do it.

What is called biological fitness is the coin of the realm by which we measure the evolutionary success of an individual or a species. To be fit, an individual needs to survive. But that's not enough. Fitness is estimated by how many offspring – actually, how many fit offspring – an individual leaves. I've put some of my resources into surviving – not getting eaten, finding food and a place to settle down, and so on. But reproduction also plays a big part in my overall fitness. To be biologically fit I need to have lots of offspring, who must survive and be fit themselves.

In birds clutch size is constrained by the amount of food the parents can bring to the nest. For robins optimal is about four. Ducks often range up to a dozen or more, because shortly after hatching the ducks become self-sufficient. So it would be fitness-folly for a robin to try to raise a dozen young, and a duck only four. As you can see, this fitness thing is complicated. But all other things being equal (remembering that of course they never are!) I want to produce lots of gametes – eggs or sperm. If I'm a male, I might also be willing to provide my mate with resources to either make more eggs or, if parental care is part of the picture, take care of the copies of my genes we've made in our offspring. More about that, below. But I should also be careful about who I mate with, not wasting my reproductive resources – eggs or sperm – on a mate who's not going to contribute to the number or fitness of our offspring. So biologists have identified a special kind of evolution called sexual selection, which occurs whenever there is mate choice: one individual deciding who to mate with. Nature is full of *Meet Markets*. And at least until blood alcohol levels rise above a certain level, similar careful mate-choices are being made in bars and restaurants from Cancun to Calcutta.

Sometimes sexual selection and natural selection are at logger-heads. A trait that attracts a mate, after all, might also attract a predator. Think peacock. Or might make it just plain hard to get around. Think elk, or rhinoceros beetle. And there's always a gamble involved in who you choose to put your resources into. Pity the poor bloke who blows his or her paycheck on the big date that turns out to be a bust.

Because of the importance of reproductive behavior on overall fitness, biologists spend a lot of time studying reproduction. So it's not (only) out of an obsession with sex that we know so much about the sex habits of insects and other creatures, animal and plant alike.

Insect sex can be rough, and dangerous. As they struggle to mate, the female of some species sometimes ends up with a half-dozen holes in her head from the male's clasping appendages. But not having sex can also be dangerous. Male *Tribolium* beetles kept from access to females live only about a third as long as those who do have access, most of the virgin males dying from their overflowing seminal fluids solidifying and sealing off the tip of their abdomen.[53] The urge can be strong. Males of the Australian beetle *Julodimorpha bakewelli* have been known to mistake a discarded beer bottle for a female, attempt to mate with it, and in the act be so determined he doesn't see approaching predators.[54] It's not funny, girls.

Much has been written about the reproductive behavior of our species, and others. Let's get back on track and focus on that part of the monarch's life. It may not be particularly titillating, though it can sometimes be downright disgusting.

There are two distinct kinds of mating behavior in monarchs – that which occurs in summer-breeding individuals and that which occurs among the over-wintering clusters of California and Mexico. The courtship behavior of summer-breeding monarchs is most of the time not particularly social. As with humans, it is usually a fairly private affair. Like us, monarchs make choices about where they hang out and feed. It appears that females in Australia choose to hang out in a particular milkweed patch depending on how many males are around.[55] Fly in, find some guys, stay. Fly in, only other girls, look elsewhere. It's the logic of the *Meet Market*.

Female checkered white butterflies have been seen chasing males, a kind of solicitation which may lead to mating.[56] (Behavior, of course, which human females would never engage in.) Female monarchs do not pursue males. As we're about to see, they're more often running in the other direction!

After years of observing monarchs in Australia, Myron Zalucki noted several things. For one, he noticed that more eggs are usually found on isolated milkweed plants than on plants growing in a patch. That seemed odd. If a female wanted to lay as many eggs as possible, as quickly as possible, you'd expect her to oviposit in patches.

But Zalucki also noticed that males tend to patrol and perch around edges of a patch. It seems they are using the patch as a kind of bait for females. The males investigate just about anything that moves. If they encounter another male, they chase him away with an agonistic spiral-flight, then return to the patch.

The Red Admiral butterfly, *Vannessa atalanta*, is well-known for this kind of aggressive male territoriality. I've sat of a quiet mid-summer's evening and watched them in my backyard, as they chase intruders away in a frantic, upward spiral flight. The males were so intent on keeping an eye on their arena they would land and rest on me between various bouts. Some of them were all beat up, their scales so rubbed off they'd lost almost all their color, their wings tattered, like veteran Sopwith Camels after too many dogfights. It seemed a miracle they could even fly. Worn and tattered wings on a butterfly, by the way, are only a rough indication of age or mileage. One good hard night in a thunderstorm, or an attempt to fly through a hard wind or across a busy highway can age a butterfly quickly.[57]

While the male monarchs are surveying their patch, any female that happens to come by gets pursued. Sometimes this ends in mating, but many females escape by doing a power-dive, or dodging through vegetation. This suggested to Zalucki that the reason fewer eggs were found in patches was because these aggressive males were actually keeping females out of the patches. He hypothesized that if he removed the males from the patches the number of females per patch would increase. Contrary to what he expected, the number of females in patches without any males went down. In monarchs, multiple mating by the female (polyandry) is much more common than in other butterflies. Patches have a male-biased sex ratio, and for those females interested in re-mating, are a good place to find males – a monarch *Meet Market*. But because the males keep bothering them at the patch, Zalucki reasoned that once mated they deposit their eggs elsewhere.[58]

There may be another dynamic here. Just as crowded railway stations and trams attract pickpockets, predators looking for monarch larvae or eggs to feast on would likely search patches of milkweed where, according to this logic, there would be better picking. Zalucki and Kitching[59] found higher mortality among monarch larvae in larger milkweed patches, perhaps for that reason.

Monarchs, by the way, are one of the most promiscuous and sexually-active species of butterflies. Both males and females mate on average about 8 times during their short lifespan, while for other species the average is about 1.4 matings per lifetime (Dr. Chip Taylor, Dplex-l list.) Dr. Taylor observed one male mating 19 days in a row, and a female with 14 spermatophores meaning she'd mated 14 times.

In many insects the sense of smell plays a big part in courtship. One sex notifies the other of its presence and availability through chemicals they release into the air, called pheromones. Research suggests we humans may, without our being conscious of it, partake in similar behavior. Pheromones play a part, though as we shall see, a relatively small part, in the courtship behavior of monarchs.

Male monarchs tend to be slightly larger than females. But they can most easily be sexed by the alar patch only males possess along a vein on their hind wings. Danaid butterflies possess specialized "hairpencils" which they extrude from the tip of their abdomen and rub into the alar patches, like a buck rubbing its scent glands in a tree. Each hairpencil has about 400 minute hairs on it. The hairpencils emit odor chemicals — at least 47 different compounds. Exactly what role the alar patches play in our monarch's courtship is as yet unknown. Unlike most male danaids, the male monarch seems not to transfer pheromone to his alar patch. Instead he waggles his tail in the air, releasing the pheromone molecules.

Though pheromone-mediated courtship is less important in the monarch than in other danaids, its close relative the queen butterfly (*Danaus chrysippus*) has been extensively studied and deserves a small sidebar to the true monarch's story. Male queens actually anoint the female with pheromones. They go so far as to produce a chemical (known to be in a class of chemicals called ketones) which inhibits flight in the female, and a glue that adheres that chemical onto the female's antennae.[60]

Male queens have been observed in the wilds of Kenya and in the lab sucking extensively on *Heliotropium* plants, which contain a precursor of the male pheromone. Males raised indoors on milkweed as their only food lacked the pheromone known from the hairpencils of field-caught butterflies. If the lab-reared males were allowed to suck on the *Heliotropum* or extracts of it, they developed normal amounts of the pheromone. Using radio-labeled precursors, researchers showed that the pheromone, (E,E)-3,7-dimethyldeca-2,6-diene-1,10-diolic acid, is synthesized within the males from these plant precursors.[61]

Components of the male danaid pheromone include a molecule called dihydropyrrolizine, also called danaidone, and several related compounds. It is these chemicals which the butterflies cannot synthesize completely on their own, instead requiring raw materials found in their food. These pheromones have been found in most male danaids but not in the monarch. In fact both the alar organs and the hairpencils are present, but much smaller in monarchs than in most danaids, suggesting chemical courtship is less important than in other related species.[62]

In some species of butterflies males are somehow able to detect the sex of the pupa while still in the chrysalis, lay claim to it as their own, and will actually fight over the chrysalis in order to be the first to mate with the female the moment she emerges. Summer monarchs on the other hand usually mate when they are three to eight days of age. Shortly after mating the females begin laying eggs. As mentioned above, monarchs are definitely not monogamous; both sexes commonly mate with several different individuals.

Instead of a sensual chemical courtship the male monarch often uses coercion, violent enough that some biologists label it as rape. Females are often seen struggling to resist the advances of males, who are very persistent. The actual mating can occur on the ground, in the air, or in trees.[63] As far as we know, among all butterflies only in monarchs are the males able to force unwilling females to mate. Robert Pyle (*Chasing Monarchs*) describes monarch mating behavior in none-too-complimentary terms. He writes that the male pursues the female, nudging her in the air, driving her to the ground, then wrestling her into submission. In some cases he's been seen to hold her down on her back, with her wings spread out, and assumes a face-to-face position. On the rear of his abdomen he has a special clasping organ with which he grabs her abdomen. He then inserts his... aedeagus.

Thought I might slip off the PG-13 shelf didn't you? Well, I'm not done yet! Firmly in his grip and aedeagus in place, he carries her with him on a nuptial flight, usually settling into a tree where they remain *in flagrante* for hours, giving him time to transfer his spermatophore into her...bursa copulatrix.

Aedeagi, by the way, are just those genital organs which entomologists often spend hours peering down at through a dissecting microscope. They are the organs John Burns, then President of the Lepidopterist Society, referred to in his 1996 Presidential address "On the Beauties, Uses, Variation, and Handling of Genitalia." In the address he states: "From the beginning I have handled genitalia in a novel but expedient manner..." OK, John. Enough!

Particularly energetic male monarchs by the way sometimes chase down and attempt to mate with the wrong species or other male monarchs, these male-male mating attempts occurring as often as a third of the time. It's not funny, girls.

Most often successful mating occurs in the afternoon, and actual copulation almost always continues until after dark, when sperm transfer occurs. The average duration of copulation is about 9 hours, but for pairs held in constant light, that increases to 27 hours. It is not known why nightfall is the cue for sperm transfer, though perhaps it is because during mating the pair is more susceptible to predators.[64]

Monarch mating behavior is unusual for butterflies, being more brutal and less dependent on pheromones. But pheromones do play a part. Male bird song has been called an auditory narcotic; it lulls the female into receptivity. In the queen butterfly, discussed above, the male pheromone seems to do just the same. A passing female, as though unable to control herself, drops from the sky and alights. He swabs her with more pheromone — or waves it around in front of her antennae — then mates with her. We speak of bouquets of flowers — and use them when courting — as well as the bouquets of wine, a sometime courtship accoutrement. You could call male butterfly pheromones aphrodisiacs. But they also act as recognition chemicals. Females can be assured that males who emit the proper mix of pheromones belong to just the right species. In the equation of fitness, mating with the wrong species is a big mistake, since no offspring arise from such a mating.

Is there female mate choice in monarch mating behavior? If so, what qualities are the females looking for in an attractive male? One study using digital imaging of male wing-color found that males with less-saturated orange coloration mated more often than those with more orange wings. There was also evidence that males with larger wings had more matings.

While transferring sperm to the female in some butterflies the male presents her with what scientists call a nuptial gift. Like a bouquet of flowers or fine jewelry, or a special bottle of wine, there's a subtext to the giving, and to the receiving. Nuptial gifts are common in insects, though in the monarch they appear to be limited to the package of sperm with its accompanying spermatophore. The male preying mantis, and some spiders (think black widow), donate their very body to the female during mating, to guarantee she has enough nutrients to make eggs for his sperm.

The evolution of gift-giving raises interesting questions. What type of population characteristics and reproductive behavior are more likely to give rise to gift-giving behavior? And once the behavior evolved, how would it affect mating behavior? It appears that in gift-giving butterfly species the females are most often polyandrous, mating with more than one male, and in these species the size of male and female individuals seems to vary more than in non-gifting species. Since the females use a combination of their own resources and the gifts to produce eggs, smaller females who have fewer resources to make eggs would theoretically need more gifts, and might therefore be expected to mate more often. This would lead to evolutionary pressure for the female to mature earlier, to allow her more time for more matings. Surveys of a number of different species support these inferences.[65]

In the polyandrous butterfly species *Pieris napi*, virgin males transfer a large nuptial gift, which is about 14% nitrogen dry weight, the equivalent of about 70 eggs. Females use this nitrogen to make their eggs. But there is conflicting evidence about the usefulness of this nitrogen. One study suggested that the number of matings – and amount of nitrogen donated by the males – did not affect female longevity, total number of eggs laid or individual egg weight.[66] A later study found just the opposite. As of now it's not possible to say exactly what part male vs. female-derived resources play in egg production.[67]

Male monarchs transfer between 5 and 10% of their body mass with each mating. The spermatophore is mostly water and protein, which makes up about 20% of the dry mass.[68] Male monarchs slip their nuptial gifts into a special "purse" possessed by the female, called the bursa copulatrix. Detailed dissection of this part of the female's reproductive tract has revealed its structure. The tiny spermatophores containing the sperm come to rest in the bursa copulatrix. On either side of the spermatophores is a row of hard teeth made from chitin, the complex polysaccharide which makes up the hard shells of insects, lobsters and crabs. Above and below the spermatophores are plates made of the same material, each plate covered with bristles. When the bristles sense the presence of the spermatophores, muscles attached to the chitinous teeth begin contracting. As the teeth grind away at the thick walls of the spermatophore, the sperm are set free to fertilize the female's eggs.[69]

Females who've mated five times were found to have ejaculates which averaged in total 38% of what the female weighed when she crawled out of the chrysalis.[70]

Whether overall reproductive lifespan is correlated with the amount of male-derived nutrients or not, it is correlated with a female's size, larger females laying more eggs. One study found that lifetime fecundity was highest for females who received the largest first spermatophores. When they died, females were usually found to contain unlaid eggs, and still weighed on average about 88% of what they weighed as they crawled out of their chrysalis. Their average egg-laying lifespan varied from 22.5 to 28.7 days. Those that lived longer, naturally, laid more eggs. This suggests that total female fecundity – how many eggs they lay in their lifespan – is a combination of how much food they got as larvae (which affects their adult size) and how many nutrients they get from their first nuptial gifts. How much nectar they get as adults doesn't seem to play a role in their overall fecundity. Nectaring, it seems, keeps them alive and flying, but doesn't contribute to their eggs.[71]

The tendency of females to mate with more than one male affects male characteristics, too. One study in the Pieridae and Satyridae, another group of butterflies, showed that the relative mass of ejaculates and the ability of the male to produce sperm and nuptial gifts was higher in those species in which the females mated with more males. In monandrous species of butterflies, in which the female only mates with one male, the male's first ejaculate is significantly larger than later ones – after all, she's not going to be mating again, so you give her as much as you can. But in polyandrous species, the male's ejaculate stays the same size over a number of matings, so long as he's given enough time between matings to build up a nuptial gift.[72] This all makes evolutionary sense. If you are competing with other males who have also mated with the female, there is an advantage in producing more sperm and accessory substances, and doing so for a number of matings.[73]

In these butterflies virgin males and those who've waited longer between matings transfer more protein and larger ejaculates into the female than those who'd recently mated. Since the time it takes to completely break down large spermatophores is longer than the usual inter-mating interval for females, it turns out that any male's resources could be used to benefit another male's offspring.[74]

Some of these behavioral studies were not done on monarchs. It's dangerous to assume that one butterfly species' behavior is another's, but results from other species are at least a place to begin to make inferences about monarch behavior. Virgin female green-veined white butterflies (*Pieris napi*) are more attractive to males than those who've recently mated. In fact after mating with a female, males lose interest in her. But as time goes on since the last mating, the female again becomes more attractive. One suggestion is that the ejaculate contains pheromones which repel other males, which lasts only a while. This repellant would protect the male's investment and allow the female to go about egg-laying without being pestered by males interested in mating.[75]

In this same species, the male ejaculate can be up to 23% of his body mass. Females who were allowed to mate with males who'd recently copulated (had a smaller ejaculate) mated almost twice as often as females allowed to mate with virgin males. This suggests females have some way of estimating the total value of nuptial gifts they've been given. In order to achieve similar overall lifetime fecundities, when each gift is small they mate more often. One way of looking at this is that females are "foraging" for mating-derived nutrients.[76] Putting it that way kind of takes the romance out of it.

Since it takes some time for the male to rebuild a sizable ejaculate, and the female wants a considerable nuptial gift, is there any way for her to know if a male who shows some interest in mating has recently mated? Apparently not. Even though copulation with virgin males only takes on average two hours, and with recently-mated males almost seven hours, in a sense wasting the female's time on a small gift, in *Pereis napi* the females don't seem to have any way of telling if a male's gift will be large or small.[77]

That the male's nuptial gift does contribute nutrients allowing the female to make more eggs is suggested by studies showing that virgin females, and females who'd only recently mated had the same number of mature eggs, while those females who'd mated three or more days previously, giving her time to access the resources in the nuptial gift, had more eggs. So it appears that male nutrients may contribute to total egg production, but are not required for it.[78] However, if you want to lay as many eggs as possible, to increase your biological fitness, you'd want as many egg-producing nutrients as you can get. That then would be a driving factor for the evolution of polyandry.[79]

Male monarchs deposit two kinds of sperm into the female. Some of the sperm have nucleii and are capable of fertilizing her eggs. Some don't, and are a kind of "filler," containing nutrients for the female and her eggs and acting as kind of "dummy sperm" in the sperm-competition game males are constantly playing. The "dummy sperm" will prevent another male from mating with that female, for a while. One study suggests that males deposit larger spermatophores, with more nucleated sperm, into females who'd already mated.[80] This increases their odds in the game of who gets to fertilize her eggs.

There are two general locations at which monarchs gather to over-winter. Nearly all of those born west of the Rockies make their way in a generally southern or southwestern direction to sites in Southern California or northern Baja California. A very few actually cross the divide and fly, instead, to join the great mass of eastern monarchs who over-winter in the central highlands of Mexico, and vice-versa.[81] We'll look at this fascinating phenomenon of butterfly migration in a separate chapter. But reproductive behavior is very much different at the over-wintering sites than during the summer.

Winter at these sites is long, often wet and dangerous. Many die before spring finally comes. For those who do survive, the last thing many do before heading north is to mate. Males and females have clustered together all winter. But with the warming, lengthening days of spring the hormones kick in and now they notice one another. A kind of frantic butterfly orgy results, with thousands and thousands tumbling in the hay right at the hibernation site, though some do put it off until the actual northward flight. The females carry the genes of their mates north, to deposit them in the eggs they lay. Their job done, shortly after mating many males die. But Chip Taylor (Dplex-l list) reports that even as far north as Douglas County, Kansas, many spring migrants were males, apparently traveling north with the females and mating as they went. The question of the differential survival of males or females while over-wintering and migrating north appears to deserve further study.

Pismo Beach is one of the larger over-wintering sites in California. By spring, males seem to outnumber females there, and compete with one another for them. The males behave as crudely as their summer counterparts, pursuing females, pouncing on them while they're roosting, sometimes overtaking them in flight and wrestling them to the ground.

Spring females show evidence of the wear and tear of this male rudeness, especially the smaller females who have to fight off larger males. Smaller females, unable to fight the males off, often mate in the morning. There's a distinct disadvantage to this. Since mating goes on until dark, they remain attached to the male all day and don't have time to forage or drink. Larger females, it has been found, keep the males at bay until the afternoon, so they can at least spend the morning tanking up on nectar and water. As the spring progresses, and the northward diaspora approaches, mating by the males appears to be more random. Since there were fewer females left, the males seem to have adopted the strategy of mating with any available female.[82] Remind you of the *Meet Market*?

As for the males, smaller individuals tend to breed earlier in the season than larger males. This may occur because the smaller males have to fight harder to get a female to mate, their wings tend to be more worn, and less capable of carrying them northward. To them the best strategy just might be to mate on the over-wintering site or not at all. The price they pay for this, though, is a steep one: the nutrients they provide a female, with their nuptial gift, could help her produce fertile eggs from sperm she's gotten from a later mating.[83]

One theory about why monarchs gather into great clumps and clusters during the winter is that it's not so much about surviving the winter in an appropriate climate, but rather about gathering together to mate. The over-wintering sites, where millions of butterflies gather, are by this theory a kind of transcontinental lek. A *Meet Market* at the Mexican *Club Mediterranea*. Others say the clustering behavior, in which clumps become so large they bend down and sometimes break sizable branches of fir trees, is to protect them from freezing. Brower *et al* used microprobes inside and outside of clusters in Mexico and noted warmer temperatures inside the clusters during the night and cooler during the day, and higher relative humidity, conserving precious water.[84] Perhaps clustering is a response to deteriorating environmental conditions, but California populations of monarchs also cluster, at sites where freezing temperatures are very rare. California over-wintering females, unlike those in Mexico, have also been seen to leave the cluster at times and lay eggs (Mona Miller, Dplex-1 list.)

As spring comes to the mountains of Mexico, the warming butterflies use up their dwindling fat supplies more and more rapidly. Females who wish to fly northward, and have enough energy to make eggs, get that energy from male spermatophores. By studying lipid levels, and dissecting females mated with radioactively-labeled males, researchers showed that female lipid levels increase after spermatophore transfer. The lipids transferred by the male were seen to make their way first through the female reproductive tract and then throughout the abdomen. A computer simulation predicted that over-wintering monarch populations would go extinct in less than a century if multiple matings by the female did not provide them with the nutrients necessary for egg production. The multiple matings, these biologists argue, are made possible by over-wintering in aggregates. Multiple mating, by providing them with lipids, may also be a significant factor in female survivorship in summer populations.[85]

Monarchs mate outside the butterfly box. That is, their mating behavior relies less on subtle aphrodisiacs and more on male coercion than most species. Why is this? Oberhauser and Frey[86] suggest that coercive mating evolved in the post-over-wintering individuals. For them, their lipids pretty well used up and the day of reckoning not far over the horizon, the urge to mate and get their DNA into future generations is especially strong. Males under these hurried, stressful conditions, are less selective than they might otherwise be – male to male copulation attempts are quite common. About three-fourths of the time females actually do fight off the coercive males at the over-wintering sites, using a variety of behaviors.[87] But they likely have less to lose from a little extra nutrient in their bursa copulatrix than they do from continuing that fight. The authors suggest it is this post-over-wintering combination of randy males and less-selective females which might explain monarch mating behavior, carried over even into the summer.

Okay. The morning after. The fun is over. The work begins. You're a fresh young she-monarch, recently mated, on a summer breeding grounds or just off hibernation. Time to lay some eggs. But not just anywhere. First you have to find some juicy milkweeds. Larvae born of eggs laid anywhere else will starve.

Fortunately you have an aerial view from which to search for milkweed. Swooping down, as you get closer, you use your antennae to smell a plant. Milkweed, yes. So, you land. But before depositing one of your precious eggs, best to double-check. Now you walk the walk across the leaf, tapping it gently with your fore-tarsi and drumming on it with your mid-tarsi and then your antennae. All these limbs have contact chemoreceptors which allow you to taste the leaf every time you touch it. You prefer a leaf with just the right amount of glycosides, and a young leaf, richer in nutrients than an older leaf. Your chemoreceptor cells tell you if you've got it right.[88]

OK, Houston, we have a go. One last thing. Has anyone already set an egg on this plant? If so, your little one will have to compete for the leaf, not a good thing. So with your antenna you also sniff for oviposition pheromones, left by another female who laid claim to this plant as her own. All clear. Slip on over the edge, touch the ovipositor, slide out an egg, launch yourself and head for another plant. One egg for each plant, one new monarch whose kingdom will stretch to where the green horizon falls off at the end of the world.

One egg per plant means one larva; your gift to your child is the entire plant you "plant" him or her on. It's sensible, and the most common egg-laying strategy. But there's another side to the story. Carol Cullar (Dplex-l list) noted that while one egg per plant might be the rule in the northern U.S., in Texas both in the spring and fall it's quite common to find a dozen or more eggs per plant. Some of this is due to what is referred to as egg-dumping. It seems that some females having made their way north arrive at a patch of milkweed and can't stop themselves, depositing twenty or more eggs on a plant. They'll also do this in captivity, when there aren't enough milkweeds around to spread the eggs onto. It's not smart, but perhaps they can't help themselves. Karen Oberhauser (Dplex-l list) notes that older females, who do not have the wherewithal to make longer flights, might also be egg-dumping. It has been observed that when more than one larvae find themselves on the same plant, they've been seen to fight over it, and in some cases even kill one another. Like the two eaglets who hatch out of eggs, the stronger to push the weaker out of the nest, this is a kind of natural selection in action.

As a larva you learned what a milkweed tastes like, while munching it day after day. There are two slightly different common species of milkweed east and west of the Rockies. Provided the choice, you prefer to oviposit on the species you grew up on, even though your larvae have an equally good chance of developing on either species.[89]

You'll lay hundreds of eggs, but only a very few will survive to adulthood. After all, the eggs themselves are protein-rich morsels for chickadees and other birds. That, and to keep them from drying out in the hot sun, is why you nearly always lay them on the bottom of leaves. The newly-hatched first instar larva is vulnerable, too. Even the Asian lady beetle, the new kid on the block, likes baby monarchs. First-instar larvae are sun-aversive. You can demonstrate this yourself. Pick a milkweed leaf with an egg on it. Keep it til the tiny larva hatches. Put it in the sun. In a little while it'll have found its way to the opposite side of the leaf. Turn the leaf over. The larva hustles off to the shade. They're still vulnerable, but as they grow, the larvae gather bad-tasting poisons into their bodies and become less desirable. Still, as previously mentioned, only one in ten or fewer are likely to reach the chrysalis stage.

We all run up against resource limits. I want to buy that camping van and a new kayak for next summer's fun. My son, meanwhile, emails me his latest tuition bill. I surf for cheap flights to Mexico while struggling to put a little more away for retirement. Time too is a resource. I'd like to go to that concert Saturday night, and the new movie that's just out; let's see, there's wash to do, it'd be a great afternoon for a hike, and my brother wants me to go on a motorcycle ride. Oops I forgot about the exams that need grading.

Butterflies don't get tuition bills, or buy boy toys. But they deal with similar decisions, though many of theirs are already made for them by the hard lessons of natural selection. Those choices (such as finding shade on the underside of the leaf) are hard-wired into their genes.

As for material needs, without a retirement fund to buck up or a checkbook to balance, monarch materialism focuses on calories and nutrients, and what to do with them. Proteins, like discretionary dollars in my own wallet, are often in short supply. How to spend those proteins takes some smart deciding.

If you're thinking of breakfast, eggs are a good source of protein. But if laying them is what you have in mind they're a protein sink. So as much as Missus Monarch would like to lay an infinite number of eggs (to improve her chances of her genes passing on), well, her protein checkbook isn't that flush. Every egg is a debit. So how many eggs should she lay? If she lays too many, they'll all be small, without enough yolk to carry the baby monarch inside the egg to hatching time. So the game plan is to lay as many eggs as she can, given the protein she can afford, but put enough protein-rich yolk in each egg to insure its survival. Turns out that's somewhere between two and five-hundred eggs, though one captive female – presumably very well fed — laid a record of 1179.[90]

We've focused on mating and egg-laying as strategies for optimizing fitness. But to mate and lay eggs requires first of all that each monarch survive. The Horsemen of the Butterfly Apocalypse ride different mounts than their human counterparts. Predators like mantids and birds are one source of mortality. More and more cars speed down more and more highways, which for one reason or another butterflies must cross – more on this in the last chapter. There are parasites to overcome, and diseases which run through monarch populations as they do any species. You have to find nectar to fuel your cell's fires, or you'll starve. Monarchs have to commit some resources to surviving all these threats before even considering reproducing.

While scientists have learned much about the reproductive biology of monarchs, there is still a lot to learn about other aspects of their biology. We'll look at what we do know, in the next several chapters.

Chapter 3
Ecology

Some words are new. *Google*. Some are very old. *Sun*. The word *ecology*, coined in 1860 by the biologist Ernst Haeckel, is but a brash teenager. Haeckel used it to describe the interplay of an individual (or species) with other individuals or species and with the non-living physical environment. It's root is the Greek word *oikos*, or household. Eco-nomics is the keeping of the "household" budget. The botanist Theophrastus, Aristotle's student, had used oikos in an ecological sense 2300 years before Haeckel when he argued that eating meat was unethical because animals belonged to the same household as we do. Philosophers back then argued over such things – whether it was ethical or not to eat meat. How quaint.

Darwin's *Origin of Species* was published in 1859, just one year beford Haeckel coined the word "ecology." Evolution and ecology share similar birth-dates, and are related by much more than that. The great American ecologist and polymath George Evelyn Hutchinson created a wonderful metaphor for their interplay: *The ecological theater and the evolutionary play*. Ecology is the stage and the list of characters while the plot, which happens in time, is evolutionary.

When you think about the ecology of a species, you're really thinking about every aspect of its way of life – what it eats, what eats it, how it stays warm or cold, where it hangs out, it's reproductive behavior and so on. A Midwestern chickadee is infinitely better at doing all that than an ecologist is at understanding it. But the chickadee, so far as we know, does not understand how one year's failure of vole populations in Canada can bring terror down from the north in the form of hungry sharp-shins and goshawks. Ecologists revel in making such connections. As generalists, we love to put puzzle-pieces together. But we are also scientists, and science's greatest strength – and some say it's greatest weakness – is in taking the puzzle apart and studying the pieces one by one. So most scientists are specialists – our professional journal articles can be really quite incomprehensible to all but a few of our fellow colleagues.

An important aspect of a species' way of life is its feeding strategies. Like some scientists, some species are generalists – they'll eat a wide range of foods. Black bears are a good example. Like humans, they are omnivores and eat everything from leaves, roots, and berries to insects, chipmunks and young deer. Other species specialize. Pandas are a kind of bear that has specialized on eating bamboo. It eats little of anything else. Specializing can bring benefits. Think neurosurgeon. But specializing has its dangers, too. Think typewriter repair. In the ever-changing natural world, these dangers can be very real. If you depend on one kind of plant for food, or a plant depends on you for pollination, you've linked your well-being to theirs. If they prosper, you prosper. If they dwindle, you dwindle. If they become extinct, you do too. Some species of tropical orchids have evolved to be pollinated by only one species of bee or wasp. If the pollinators become rare, the orchids do not produce seeds, and their populations dwindle, bringing bee populations down, and so on. The dynamics enter a downward spiral.

The "marriage contract" two species enter into when they agree to share a future is called coevolution. In Shakespeare's *Macbeth*, Macbeth and Lady Macbeth commit themselves to an evolutionary future which in time takes turns they'd not calculated. Their coevolution leads to both their extinctions. Hamlet's tragedy is of a different sort – unsure of who to trust, he becomes stuck in the mere considering of which evolutionary path to take, and becomes in a way, like the chickadee caught in the goshawk's talons, the victim of a fate he cannot quite comprehend. Perhaps that's why we empathize so readily with him – and with the chickadee!

Now that we know where the word "ecology" came from, it might also be a good time to ask from where came the name "monarch" for this species of butterfly? Monarch butterflies were named, in a roundabout way, for another monarch, who was born Prince William of Orange-Nassau, in 1650. Forty years later by defeating his uncle and father-in-law King James II of England, at the Battle of the Boyne, he became King William III. As king he founded a college in colonial Virginia which he named after himself and his queen Mary. In Scotland his order to extirpate the clan MacDonald resulted in the brutal massacre at Glencoe. For his sake the Irish flag is one-third orange, and each July in Belfast and Dublin the Orangemen parade.

While Prince William fought for the throne in the revolution of 1689 his followers signified their loyalty to his cause by wearing orange-colored ribbons and scarves. In the American colonies they named a big orange and black butterfly, not found in England, for him: King Billy. Those less partisan gave the butterflies a more neutral name: monarchs.

Was the King Billy for whom the butterfly monarch was named infected with Hamlet's disease? The evidence suggests not. He boldly declared himself enemy of his uncle James II and devil take the hindmost. Then by wearing the orange symbol of loyalty his followers committed themselves to "coevolve" with him.

King Billys, they were called. Then, monarchs, the name that stuck.

As we've already seen, our butterfly monarch has entered into a very committed relationship with milkweed plants. The monarch larvae will eat almost nothing else; their health and growth depend on milkweed. But because adult monarchs nectar on many different plants, the contract has a closing date – their entrance into the chrysalis. Because of the larva's near-complete reliance on the milkweed for food, if a virus spread rapidly around the world and killed all milkweeds, the monarch would quickly follow it to extinction. This is a one-sided relationship, by the way. Since many other insects pollinate milkweed flowers,[1] the milkweed is not dependent on the monarch, and if monarchs went extinct we can imagine milkweeds barely batting an eye.

What do we know about the history of this relationship? It has definitely been a long-term commitment – hundreds of thousands of years or more. There are about 2400 species of milkweeds in the world, all belonging to the family Asclepiadaceae. Evidence suggests that the genus *Asclepius*, with which monarchs have coevolved, originated 40-50 million years ago in the Caribbean, Central America or very northern South America.[2]

The genus *Danaus*, to which our monarchs belong, came from the same home-town as milkweeds and appears to be about five million years old – only one-tenth as old as the milkweed. Though they have relatives in Africa and Asia, all evidence suggests our American monarchs (*Danaus plexippus*) evolved in the western hemisphere, perhaps only a quarter million years ago. So unless someone brought him a monarch from the colonies, England's King Billy would never have seen one! Storms and strong winds sometimes carry birds to places they're not expected to be found. It's presumably how, long ago, birds and insects first reached the Galapagos, and other scattered islands. Monarchs have occasionally been spotted in Newfoundland, where milkweed isn't found, presumably brought north by storms (Don Davis, Dplex-l list). Only in the last hundred years or so, as our species increasingly globalizes the world, has *D. plexippus* hitched longer rides, from continent to continent, showing up in the Azores, Canary Islands, Portugal, Great Britain, France, Africa, Hawaii and Australia.

Here's a kind of genealogy of the two royal families – the two King Billys:

England's	King Billy	Monarch butterflies
Domain	Eucarya	Eucarya
Kingdom	Animalia	Animalia
Phylum	Chordata	Arthropoda
Class	Mammalia	Insecta
Order	Primates	Lepidoptera
Family	Hominidae	Nymphalidae
Genus	*Homo*	*Danaus*
Species	*sapiens*	*plexippus*

As you can see both kinds of monarchs — humans and the insect kind — are eukaryotes and animals. We are eukaryotes because we have nuclei and mitochondria inside our cells. As animals we store our excess sugars as glycogen rather than starch, and don't have cell walls. But after that, our evolutionary genealogies make us quite different. As chordates we humans have a backbone, as mammals hair and mammary glands, and the trappings that come with being a primate, such as good eyesight and five fingers with fingernails. We're the only surviving species in our genus — our ancestors *Homo habilis* and *Homo erectus* are extinct, as are our cousins *Homo neanderthalensis*. We are *Homo sapiens*, the intelligent hominids. We of course had the advantage of naming ourselves.

Butterfly monarchs are not chordates. Being arthropods, instead of backbones they have exoskeletons. Their six jointed legs and three main body parts as adults— head, thorax and abdomen — make them insects. Their scaly wings put them in the order Lepidoptera — that's what Lepidoptera means – *ptera* are wings, and *lepido* means scaly. (The bacterial infection which causes leprosy is so named from one of its symptoms – scaly skin.) Monarchs belong to the genus *Danaus*, species *plexippus*. They were given that name in 1748 by the great biologist Carolus Linnaeus, who coined our two-name (binomial) system of scientific nomenclature. Because of their migratory habits the European species which Linneaus encountered were sometimes called "wanderers." Linneaus knew his classical mythology – he named many species for one or another mythological character, who after all could not complain about sharing their name with a reptile or insect.

King Billy was a real prince. There was another prince, a legendary prince of Egypt named Danaus. As prince, he had many wives and many children, including some 50 daughters. In those ancient days without CNN or weather.com, princes kept fortune-tellers close at hand. Danaus wondered how he would die. He asked his fortune-teller. You will die, the fortune-teller warned him, at the hand of a son-in-law. So began Danaus' journey through logic and across the world. If my daughters don't marry, I'll have no son-in-laws, he reasoned. So how do I stop them from marrying? Well, if we constantly keep on the move their suitors will never catch up with them. It was the beginning of the Danaus family vacation. No sooner had they settled in Libya than he picked them all up and moved them to Rhodes. From there to the Peloponnesus, and on and on. For a while his plan actually worked. But daughters have ways of manipulating fathers. Dad, it's time for me to settle down and get married, they complained. Multiply that times fifty and you have to feel just a bit sorry for the poor man. So, Danaus was reduced to casting far and wide for another ruse. This one was not so subtle, and here I lose all respect for him, prince or not. He decided he would convince his daughters to kill their husbands, on their wedding-night. They all did. (Kids must have been way more obedient back then.) All but one, that is – and that son-in-law, by the name of Lynceus, eventually killed Danaus. It was his wandering life, like the monarch already known in Linneaus' time as a migratory creature, which brought Danaus to Linnaeus' mind. *Plexippus*, the name of the new- world species, appears to be from the Latin *"plexi"* meaning woven; we don't really know why Linneaus chose that name, though it might describe the monarch's proboscis, which appears as though woven of several fibers.

Taxonomy, the science of classifying organisms, is as much art as it is science. It's definitely not a career for the faint-hearted. One unfortunate side-effect of all the recent interest in "big-picture" science, from ecology to cosmology, is a waning interest in taxonomy. There are almost no graduate students in our universities on their way to becoming taxonomists — another example of a specialist flirting with extinction.

Gather a dozen taxonomists together – if you can find that many — to talk about the classification of a species and as they say, you'll get thirteen opinions. According to Ernst Mayr's biological species concept members of the same species can successfully interbreed, while members of different species cannot. Biology has its rules, but biology is a science rife with exceptions. Though uncommon, interbreeding to produce hybrids does occur.

Corals apparently are especially good at embarrassing and confounding taxonomists, casting their gametes widely and prolifically. This makes classifying corals really hard.

Above the species level – genus, order, family, and so on — taxonomy is pure inference, based as best we can on evolutionary, morphological, sometimes behavioral, and more and more on molecular evidence of who is related to or descended from whom. Specialists that they are, taxonomists have to assemble as much information as they can get into a quasi-reasonable hypothesis of who's who and what's what. As for our butterflies, most recent DNA evidence suggests the genus *Danaus* includes a dozen or so species (depending of course on which taxonomist you talk to) including *D. plexippus*.[3]

In addition to *D. plexippus*, new-world monarchs include the queen (*Danaus gilippus*), the tropic queen (*Danaus eresimus*) and the Jamaican monarch (*Danaus cleophile*), found only on the islands of Jamaica and Hispaniola. The southern monarch (*Danaus erippus*) is found in southern South America.

Other species of *Danaus* are found in many parts of the world. The Ismare tiger (*Danaus ismare*), the common tiger (*Danaus genutia*), the Malay tiger (*Danaus affinis*), and the white tiger (*Danaus melanippus*) are Asian species. The Lesser Wanderer (*Danaus petilia*), considered by some a subspecies of *D. chryssipus*, is found in Africa, Australia, and the South Pacific. The plain tiger (*Danaus chrysippus*), widespread in Asia, Africa and southern Europe, is one of the first butterflies to show up in artwork, appearing on an Egyptian fresco perhaps 5000 years ago. The Dorippus tiger (*Danaus dorippus*) is another old-world species.

In the life-history chapter we've already learned quite a bit about the feeding habits of the monarch larvae. There, our focus was on the larva itself. Now let's look a little more closely at the interaction between the larva and the milkweed; after all, interactions are what ecology is about.

When a hungry herbivore shows up for a meal, plants can't fight back or get up and run away. They have to find other defenses. Thorns, like the swords and sabers carried by King Billy's followers, are effective. Nettles, which sting with drops of formic acid, and poison ivy or poison sumac use another approach to defending the homeland. They use chemical warfare. Evolution has provided them with chemical factories for protection. This approach is widespread in the plant kingdom: many plants produce chemicals which make them taste bad or are poisonous. We humans have learned to use many of these chemicals for our own purposes. Nicotine, caffeine, cocaine, heroin, atropine, ephedrine, digitoxin, menthol, and so on are put into the plants not for our use but because they taste bad or make potential herbivores sick. Studying these chemical antifeedants is all part of a fascinating branch of ecology called chemical ecology.

Other insects than the monarch also make or sequester in their bodies bad-tasting or toxic compounds. The lubber grasshopper, America's largest, is found in the southern states. Research has shown that it is quite toxic. Loggerhead shrikes catch them and often impale them on thorny shrubs. This impaling behavior seems to serve the shrikes in several ways – the male uses it to show off his hunting prowess to females, it marks his territory, provides a ready cache of food when it's otherwise hard to find, and it also ages chemically toxic foods – like the lubber grasshopper — to make them more edible.

And monarchs aren't the only insects that use the toxins in milkweed to protect themselves. Milkweed bugs do, too. They feed on the plant's seeds, and like monarchs because they sequester noxious chemicals from the milkweed, have few predators. A genus of beetle called *Tetraopes* also appears to have co-evolved with milkweeds. A different species of this beetle specializes on almost every one of the 120 different species of *Asclepias*. Like monarchs, *Tetraope* adults come with orange and black warning coloration. The praying mantid *Tenodera ardifolia*, which sometimes preys on the milkweed bugs, throws up after consuming milkweed bugs raised on seeds of the milkweed, and quickly learns not to eat them after that, whether the bugs have been raised on milkweed or a nontoxic plant. Predacious wasp larvae have been shown to not prey on monarch caterpillars for the same reason.[4]

Though it clearly has its benefits, synthesizing chemical antifeedants is a drain on a plant's resources. To save energy, some plants have the ability to produce the antifeedants only when they're being attacked. These are called induced antifeedants. Plants which use this strategy do not waste their resources making lots of antifeedants until they are under attack. But there's more to the story. When a caterpillar or beetle chews on a leaf, volatile organic compounds are released into the air. You can simulate this by crushing almost any leaf and noticing the smell. Some species of plants appear to have evolved the ability to detect these odors as they are released by a neighboring plant that is being eaten. This allows these plants to turn on their chemical factories even before the herbivore arrives.[5] I know of no research suggesting milkweeds are capable of this.

Over time, one can imagine milkweed adapting several evolutionary strategies to counter monarch herbivory. One would be to ratchet up (induce) the production of toxins and latex. Another might be instead to put resources into repairing tissue damaged by the larvae. One study suggests that this latter is in fact the main approach milkweed seems to be taking.[6]

Milkweed plants produce a complicated cocktail of antifeedants twenty-four hours a day. But there's also evidence they do ratchet up their production if they're being chewed on. A weevil which eats milkweed induces this response, and individual plants which have been damaged by the weevil have higher than normal toxins. This makes those plants less palatable and more dangerous to monarch larvae. But by damaging the leaves, monarchs also make those individual plants more toxic for other potential herbivores. So the herbivore which first colonizes a particular plant has a lasting effect on later colonizers, and helps determine what the subsequent insect community on those plants will be.[7] Timing, as they say, is everything.

Not all kinds of milkweed have the same chemicals in them, or at the same concentrations. *Asclepias syriaca* (common milkweed) and *Asclepias speciosa*, (showy milkweed, a more westerly species) are relatively low in poisonous cardenolides, and produce less poisonous monarchs than other milkweeds. One study found 23 different cardenolides in *A. speciosa*, of which 20 are sequestered by the monarch. These cardenolides include labriformin and labriformidin, uzarigen and syriogenin.[8]

It's a bitter pill to swallow, this eating poisonous leaves to become toxic and survive. Is it worth it? In the end, how successful are our monarchs at escaping predation?

Death comes to the door of a monarch in many guises. Hard frosts. Cars on the highway. Various parasites, and diseases. Predators of many kinds want you, and not in an affectionate way. As an egg you were a rich protein morsel for birds, mites and spiders. As a larva you are on the wanted list of many predators and parasites. Tachnid flies, of which there are more than a thousand species native to the U.S. in addition to invasives, lay their eggs on the milkweed or on you. The tachnid egg hatches into a maggot which burrows into you and feeds on you. There is the protozoan parasite *Ophryocystis elektroscirrha*, discussed in more detail in a later chapter; nosema, a protozoan disease; various viruses and bacteria; and as larvae the very manna which nourishes you, because it is toxic, can kill you. As an adult many birds are all-too willing to try you out. Mantises pray for a taste of your flesh; predatory stinkbugs are on the prowl. Dragonflies watch to catch you, merlins, kestrels, red-tailed and broad-winged hawks have been seen swooping you out of the sky, and should you make yourself available to them, even wee timorous mice do not have your best interests in mind.

Predation ecologists dissect the act of predation into these components:

Encounter (meaning to find, or sense) – **capture** – **handle** — **ingest** — **digest**.

The goal of an individual predator, or in a longer-term view the species, is to be as good at each of these acts as it can be.

The goal of a potential prey is simpler: Escape! Quick! Before it's too late!

This game of eat or starve, escape or be eaten, is sometimes described as a coevolutionary arms race. Every time a predator gets better at finding, capturing, handling, ingesting or digesting a prey item, the prey species evolves to be better at evading the predator. It's a deadly game, going on around us all the time, a game which results in those remarkable adaptations so often featured on nature programs or in science museums. Whales who use their song and a ring of bubbles to corral herring. Cheetahs who can run fast enough, for short distances, to break most speed limits. Bats using a kind of ultrasound sonar to locate flying moths, and moths who can hear the bats and take evasive action, to name but a few.

In the end, it's really not about you and me, as individuals, or the individual monarch larva. That may be why evolution is such a hard pill for some to swallow. We are a culture of the individual. We are told that each of us is special. And it's true. But evolution throws us all into the sausage-grinder of statistics. Rather than any one individual's destiny, a little survival or reproductive advantage here or there is what matters in the long run, advantages measured in tenths or hundredths or less of a percentage.

After all this quiet munching of the bitter pills of chemical-saturated milkweed, the life of a monarch larva is still a dangerous life. Their mortality rate is up there in the clouds with fighter pilots in World War I. Early instar larvae, who've not yet had a chance to load up with toxins, are the most vulnerable. In one study of 882 larvae counted, only 3% survived to the third instar. The researchers watched crab spiders and ants eat these early instar larvae one after another. Some – one wants to think the brighter ones — did escape by hiding out in the milkweed flower inflorescences. Those of us who enjoy lying in the summer grass sometimes catch a glimpse of little red spider mites, making their way through what for them must be a gigantic jungle. These mites, related to ticks and spiders, consume many monarch eggs. In this study eggs or larvae didn't survive well on milkweed plants that had ants on them, presumably because the ants found them tasty.

Nearly half the fifth-instar larvae counted were parasitized by parasitic wasps. These wasps lay their eggs on the larvae which the immature wasps consume from the inside out. About 30% of the pupae, too, were parasitized by these wasps, with as many as 15 wasp larvae per pupa. Most of the parasitized pupae died. Research shows that larvae kept in enclosures, which excluded most predators, have a much better chance of survival.[9]

Science starts with an interesting observation. It is not, as my son John once pointed out to me, the archetypical Archimedean "Eureka!" This insight into how the world works comes later. The motivation to understand comes instead from the more meditative "Hmm, that's funny." One such odd observation of monarch larvae resulted in a quaint paper by the Dame Rothschild we've already met, and a colleague of hers.[10] That paper's title is: *The monarch caterpillar waves at passing hymenoptera and jet aircraft.* Out in her garden Dame Rothschild noted that loud noises – passing wasps or bumblebees, humans making buzzing sounds, shouting, singing, or even jet planes passing overhead – often caused the larvae to wave their tubercles (short antennae-like structures) around. "Hmm, that's funny." Now Dame Rothschild's innate curiosity kicked into high gear. "What's this all about?" Further observations led her to the "Eureka!" moment. The larvae, it turns out, make a noxious-smelling chemical called 3-hydroxy-2-butanone, and timorous beasties that they are, when they hear an unusual noise assume somebody's coming to do them bad. Which, by the way, is often the case. In response to the noise they wave their tubercles around, releasing their stink and, hopefully, cutting back on the capture piece of the predation equation. Try it the next time you find a monarch caterpillar. Can you get it to wave at you?

If the stink doesn't work, and the larva is captured, they pull another chemical weapon out of their arsenal. This noxious chemical, stored in a package behind their head collar, is a mix of 2-methoxy-3-alkylpyrazines, and targets the handling component of the predation equation. Pyrazines, released by other insects such as ladybird beetles, are repellant to vertebrates and invertebrates alike. They smell really bad. But not only do they smell bad, pyrazines act in other ways. They suppress the predator's immune system –

part of their nastier, long-term effects. And they seem to improve memory, deeply engraving the encounter with them in the mind of the would-be-predator.[11] These pyrazines and the related alkylpyrazines are also essential precursors to male monarch pheromones, so they are likely acting both as defensive weapons and precursors to courtship molecules. The argument has been made that the cardenolide concentration of individual monarchs is not high enough to protect them from predation, and it's the combination of cardenolides and alkylpyrazines that protects them.[12]

Let's look at another aspect of this "escape predation" thing. If you're a butterfly not protected by a reputation of being poisonous or tasting bad, and you want to escape birds, one strategy would be to fly fast in a randomly zig-zaggy hard-to-catch way. In one study in Costa Rica which looked at the flight patterns of 54 different kinds of butterflies, the nonpoisonous ones had stronger flight muscles and flew faster and in more unpredictable paths. Poisonous butterflies on the other hand tended to lumber through the air with weaker flight muscles, in straighter and slower flight.[13] They were the B-29s relying on their armament — sequestered chemicals — to protect them from swooping predators. Palatable species, without the protective chemicals, can also accelerate faster than species like monarchs, who put more of their resources into guts and ovaries than in wing muscles.[14]

It is the wings of adult monarchs which have the highest concentration of cardenolides. Since it is a piece of wing a pursuing predator is most likely to taste, this facilitates the learned rejection response. Other more emetic compounds, which cause the predator to vomit, are stored in highest concentration in the abdomen.[15] Some species of birds have learned to discard the wings and other body parts which are highest in toxins. In his book *Chasing Monarchs* Robert Pyle describes watching orioles feeding on monarchs at the El Rosario sanctuary in Mexico. Some are very picky eaters and only eat the rich fat deposits — they say the clumps of over-wintering butterflies burn like candles, a disgusting thought if there ever was one — and in that way circumvent the monarch's toxicity.

Not all birds find monarchs unpalatable. Carol Cullar (Dplex-l list) describes scissortail flycatchers in Texas feeding on migrating monarchs in this way: "A gaggle of scissortails (10-20 individuals) loosely surrounds the monarch cluster of between 100 to 300 monarchs at a distance of some 12 to 15 feet. One bird is elected to crash into the cluster, knocking between 15 to 20 monarchs to the ground below the tree, then the marauders swoop in and gobble up those on the ground. There doesn't appear to be any picking and choosing going on as to what gender of monarch to gobble! It's strictly hit and run. Then the whole show happens again until it's too dark to see."

Black-headed grosbeaks and black-backed orioles are the primary bird predators of over-wintering monarchs. One year it was estimated they'd killed nearly 10% of the over-wintering butterflies; in smaller colonies that number may have reached or exceeded half. Both species killed more butterflies on colder days. Orioles feed on butterflies in a cyclic way, building up cardenolides in their body until they begin to be toxic, then do not eat monarchs until they've cleared the toxins. Grosbeaks get around this by the way they consume the butterflies, leaving more toxic parts uneaten.[16] Great concentrations of resources – like the over-wintering clusters of monarchs – naturally attract parasites and predators. There are four species of mice in the area which could be potential predators. Of them only one — *Peromyscus melanotis* – preys on them in any significant degree. It is somehow much less sensitive to the cardenolides in the monarchs than are the other mice. It's also careful about how it eats a monarch. It rejects the cuticle surrounding the abdomen, high in toxins. Instead these mice were seen to make slits in the side of the monarch's abdomen and suck or lick out the contents. The other species of mice appear to be averse to the butterfly's bitter taste. When caged mice were given nothing to eat but monarchs, they survived on them and didn't become sick, though only *P. melanotis* was able to gain weight on a monarchs-only diet.[17]

The cardenolides protect larval monarchs in another way. We'll get to know another enemy of monarchs, the protozoan parasite *Ophryocystis elektroscirrha*, more intimately in a later chapter. This parasite infects larval monarchs. Those with high populations of this parasite die; lower numbers make them sick. Larvae who'd fed on low-cardenolide milkweed had higher mortality rates from infection by this parasite than those which fed on plants with higher concentrations of these cardenolides. This appeared to be due to the ability of the cardenolides to decrease the reproduction of the protozoans.[18]

As a scientist, to me a mystery is a goad to understanding. Though I have lived alongside them in California, we don't have rattlesnakes this far north in Wisconsin. But we do have fox snakes — pine snakes, to some — which look a lot like a rattler. I mistook some once, because when threatened, the fox snake will vibrate its tail in dead leaves. It sounds a lot like a rattler. It's classical Batesian mimicry — a harmless species copying a poisonous species. Darwin, in his journals of his trip around the world on the H.M.S. Beagle, writes of another snake, in South America, which like the fox snake doesn't have rattles but also vibrates its tail against leaves or grass. The mystery to me is how a fox snake learns to vibrate its tail to protect itself. Did its southern or western cousins pick up on the idea by watching rattlers? How could they know it was a warning and not just a nervous tic, or some kind of courtship behavior? Or did they inherit a kind of Parkinson's-of-the-tail that ended up protecting them?

In the pecking order of human nobility, monarch — from the Greek meaning the *one* authority — is pretty much on top. Next down in some societies might be the viceroy — vice-king, as it were. Just about everybody knows the story of the viceroy butterfly, a smaller mimic of the monarch. For decades the two were featured as a classic example of Batesian mimicry. Monarch tastes bad, so most of the time escapes predation. Viceroy evolves to look like monarch, and escapes predation.

There is no question birds can identify butterflies. In one study, the authors used colored markers to paint white butterflies different colors. They offered unpainted and painted butterflies to a male rufous-tailed jacamar, a tropical insectivore that naturally preys on butterflies. Unpalatable local butterflies tend to be black, orange and red, and those were the colors the bird rejected most.[19]

Florida scrub jays given a monarch tasted and rejected it, and after that never tasted a monarch again; nor did they try to eat the mimetic viceroy. They did eat non-mimetic butterflies, and if they'd never been exposed to a monarch, to a lesser degree viceroys. Three-fourths of the birds who'd once tried to eat a monarch but rejected it remembered their experience for over two weeks, and during that time stayed away from both monarchs and viceroys.[20] Viceroys from Florida, by the way, tend to resemble the darker relative of the monarch, the Florida queen; northern viceroys aren't so dark.

David Ritland [21] offered blue jays a mix of butterfly abdomens, including some viceroys, but just the abdomens. While 98% of the controls (not viceroys) were eaten, only 40% of the viceroys were. So it appears that viceroys themselves are not all that good eating. Prudic *et al* show that the viceroy larva does sequester defensive chemicals from one of its host plants, the Carolina willow.[22]

Now we introduce some new characters to thicken the plot further. In the southern U.S. the monarch (*D. plexxipus*), the Florida queen (*D. gillipus berenice*), the viceroy, (*Limenitis archippus*) and a subspecies of the viceroy are found together. The monarch and the queen are known to sequester toxins, though because it feeds on a kind of milkweed that has lower concentrations of these chemicals, the queen is not as well-protected. The viceroy has bright orange wings in the northern part of its range, where it mimics the monarch. Towards the south, the wing coloration darkens, and in this subspecies resembles the more common queen.[23]

Ritland [24] showed that the Florida queen has a wide range of palatability to red-winged blackbirds, depending on the food the larvae were raised on. Eighty-five percent of those raised on a cardenolide-free *Sarcostemma clausum* were eaten, while only 8% of those raised on *Asclepias curassivica* were. In other words, depending on what the queen larvae have eaten, the viceroy mimic might be less palatable than the queen, turning the normal view of this mimic pair topsy-turvy.

Captive red-wing blackbirds were starved for 14 hours before presenting them with the abdomens of various butterflies. The redwings readily ate every abdomen of swallowtail and red-spotted purple butterflies, about 80% of viceroys and 70% of queens. They responded to the rejected abdomens with the typical feather-ruffling and head-shaking they show for bitter-tasting monarchs. Those that did eat a viceroy didn't vomit, suggesting they taste bad but are not poisonous. Some birds would readily eat abdomens which had been rejected by other birds, so as in humans, there appears to be individual differences among the birds — some are "gourmands" and others "gourmets." Since individual butterflies feed on plants of differing cardenolide concentration, individual birds may encounter a bad-tasting or edible individual monarch, viceroy or queen. There may also be a spectrum of palatability over the course of time, as different milkweeds become more available. So Ritland suggests that in actuality queens, viceroys and monarchs are all mimicking one another, a kind of mimicry called Mullerian.

Since the butterfly doesn't munch leaves containing toxins, they eclose from the chrysalis with whatever toxins they had going into it, and over time these chemicals are slowly broken down and excreted. As they age, monarchs become less poisonous and in a sense become mimics of the younger, more poisonous individuals.[25]

An individual milkweed plant can be the stage for an ecological play whose plot is far more complex than a Puccini opera. The milkweed itself competes with other plants, sometimes other milkweed species, for resources such as sunlight, water, and nutrients. Add a monarch larva and the ecology, as we've seen, becomes fascinating. But milkweeds also harbor aphids, who've become immune to the plants' toxins, and sequester them as do the monarchs. The same for milkweed leaf beetles, whose larvae feed on the plant's roots. While it's common for ants to take care of aphids, in exchange for the sweet exudates they produce, this association is less common on milkweed. Ants might not like the toxins. If you watch a yellowish crab spider on the plant who's just eaten some of the aphids, it's web will be less symmetrical than normal. You might find tussock moth larvae, who eat the milkweed leaves even though it interferes with their growth.[26] And now, from up the road, comes a big tractor and mower, dead-set on obliterating the entire assemblage.

Monarchs first appeared in Hawaii in the mid 1800s. On Oahu there is a white morph (variety). Where the normal monarch has orange scales, the white morph has white. The patterns of black stripes are the same. This white morph appears to be the result of a founding population with a recessive trait. The white morph first caught biologists' attention when in the 1890s one was sent to Walter Rothschild, Dame Rothschild's uncle. This white morph was seen to increase from about 1% of the population in 1960 to 8% in 1988. The increase in the white morph is attributed to the accidental release, in the mid 1960s, of two species of bulbuls, birds which are immune to the monarch's chemical toxins. The species of milkweed Hawaiian monarchs feed on has white flowers, and evidence suggested the white morph blended in better on these flowers, and so were eaten less often than the orange individuals. In time, the bulbuls began to eat larvae rather than adults. The larvae of the color morphs are indistinguishable, and the abundance of the white morph has once again declined.[27] Through breeding experiments Grace Venters (Dplex-l list) has verified the recessivity of the trait. Photographs of individuals of the stunning white form, raised by Don Wilks, can be found at: http://www.monarch.org.nz/monarch/2008/07/27/more-white-monarch-photographs/

There's an interesting and little-known sidebar to monarch mortality. We are all of us mostly not-us. If we are fifty trillion of our own cells – a pretty good guess — we are perhaps ten times that many much, much smaller bacterial cells, most of them in our gut but scattered all over us. Billions in our mouth, for example, of five hundred or more different kinds. They've taken up home all over our skin, in our mucus membranes and elsewhere. As individuals we are all of us walking, talking microbial ecosystems.

Of course the same is true for insects. One well-known example is the bacterial community in termites' guts which helps them digest wood. To assure that young termites aren't born without them, some of these bacteria are packed into each egg-case by the queen. There are other examples of beneficial bacteria being packaged into eggs.

But there's an interesting twist to the story. When the female of a number of different kinds of insects deposits an egg, it might be infected with a bacterium that kills only the male offspring. It's not a hundred percent effective, but most males die. Eggs destined to grow into females might also be infected, but the infection doesn't kill them. Two of the bacteria implicated in these nefarious deeds are species of *Rickettsia* and *Wolbachia*. It's not known how this bacterium kills only males, or whether as a female lays an egg she can choose to infect an egg or not. *Danaus chrysippus*, an African species, plays this dangerous game. It appears that something like 40% of the females are infected with this bacterium, which is closely related to two ladybird beetle male-killers, and to the wood-tick symbiont *Spiroplasma ixodetis* which causes lyme's disease in humans.[28] It is not known if any of our monarch, *D. plexippus*, are infected with male-killing bacteria.

What is this odd story all about?

In any species, females are really the reproductive members of the population; males are only there to make sex work. That is, they provide the population with the genetic diversity necessary for evolution. In some cases, such as honeybees, the males are haploid, carrying only one copy of each chromosome rather than two. In this way the population purges itself of harmful recessive traits – males carrying them die without reproducing.

Some species of lizards are made up of females only. Zooplankton like *Daphnia* will produce millions of females and only produce males under deteriorating environmental conditions when sexual reproduction and the genetic diversity it brings might be a good thing. Normally the population gets along just fine without males, who compete with the females for limited resources. In some insects, the males killed by their inherited bacteria are consumed by the females, becoming a population resource rather than sink. So the ability to control the proportion of males in the population is, for the females at least – and after all, they do control the "egg-strings" — an evolutionary advantage. It's not funny, girls.

Chapter 4
Migration

Butterflies had played their earnest games with the air's fickle currents for millions of years before our species stood up to look around, liked what we saw and proceeded to claim it as ours. Long before they'd discovered the uses – and pleasures — of language our ancestors must have marveled at these flying "dream flowers." We likely first used language to name common tools like stone knives, baskets, spears and fire, or to communicate simple requests to one another. Slowly the realm of words spread across the natural world, allowing our species not only to observe but to comment on what we saw. The words we chose to stand for what we saw can tell us as much about ourselves as about the objects they point to.

Folk-tales, like etymologies, are fossil footprints leading us back to more ancient times. These folk-tales, told around hearths and fires, in lodges, huts and tents entertained our ancestors during the long dark nights and were vehicles which carried the wisdom and ethos of their culture. Some of them speculate on the origin of the world around us, and how each creature came to be as it is. Some are cautionary tales meant to teach children about the dangers of the world or the potentially dangerous urges inside them. Romantic love, some tell us, is a burning and dangerous thing. Love stories were as popular a millennium ago as they are in Holly/Bollywood today. The story of a beautiful girl who falls in love with a nonhuman creature, sometimes a supernatural being, is found in almost all cultures.

The only surviving full-length novel from classical Rome is that written by Lucius Apuleius. He called the book *Metamorphoses*, after Ovid, but it also goes by the name *The Golden Ass*. One episode in the book tells the story of Psyche, youngest and most beautiful of a certain king's three daughters. Psyche's charm and beauty were so admired that Venus became jealous, and asked her son Cupid to cause Psyche to fall in love with some lowly, degraded creature. In *A Midsummer Night's Dream* Shakespeare adapts this story to his own use when he has Titania, queen of the fairies, fall in love with Bottom, who's been transformed into an ass. Ovid's *Metamorphoses* plays a big part in Shakespeare's play, a play full of magical transformations.

Cupid's arrow (still scribbled or tattooed by 21st century infatuated teens) pierces Psyche's heart, and takes her through many heroic adventures. She enters the same fairy-world Shakespeare takes us into, a world of magical beauty, but a dangerous world too. Embedded in Psyche's story is the

Cinderella story, of two jealous sisters who plot against a younger, more lovely sister, as is the story of the meddling mother-in-law who in this case happens to be played by Venus.

The story of what happens to Psyche is complicated and fascinating, and the seed of the better-known story of Sleeping Beauty. Cupid falls in love with her, and she with Cupid. But loving a supernatural being such as Cupid is dangerous, and she undergoes one trial after another. Finally Jupiter intervenes and grants Psyche eternal life, so she like Cupid is immortal, and the two are married. They settle down and live happily ever after. Their first child is a daughter named Pleasure.

By the 5th century B.C.E., six hundred years before Apuleius wrote his novel, the Greeks were calling butterflies by the word *psyche*, and around then butterflies began to symbolize the soul. Before Christianity turned the soul into the diaphonous ghostly thing which survives our deaths, in the pre-Christian tradition, more pragmatic and less other-worldly than the Christian, the soul was our "higher self" while still alive. In Latin it was sometimes *spiritus*, the magic of breathing, at other times referred to as our *genius*, our unique self which it is our responsibility to come to know and manifest in this world. The butterfly is to the caterpillar what the soul or spirit is to the mortal body, this clumsy eating and pooping thing which St. Francis called "Brother Ass." There is little I know of in the Christian literature about creating the connection between body and soul; they are, in this view, intercalated at birth, and separated at death. But the classical metamorphosis is a migration of the soul not from the body, but into the body; we become spiritual beings, in this view, while we are still alive. The *larvae* of the Roman tradition were the returned spirits of those who had not had time or opportunity to complete this connection before their deaths. Like the events occurring in the chrysalis, connecting soul to body is not a simple thing. It is dangerous, but it is transformative.

In a certain sense, this book belongs in the "psyche-ology" section of the library.

In Latin rather than Greek, butterflies were *papilia*, from which we have the word "pavilion," a tent with its wings stretched out. You are eating butterflies when you eat Italian *farfalla*, the wing-shaped pasta. In Spanish they are *mariposa*, probably from "Santa Maria posa," the Virgin Mary's grace coming to settle among us. Old English has them as *butterfleoge*, cousin to the German *Schmetterling*; to them butterflies were thought to be little fairies or witches who come to steal milk and cream. Surely milk was spilled on ancient farms, or dripped sometimes from full un-milked udders, and just as surely nearby butterflies flew in to lap it up. The Germans have a second name for butterflies, *Tagfalter*. Though some consider *Tagfalter* to mean "day-flyer," it's real etymology is more likely the less obvious "day-folders." And yes, Teutonic logic then gives us moths as....*Nachtfalter*. Norwegians call them *sommerfugl*, summer fliers. In Russian they are *babochka,* because they resemble the knots on scarves worn by old women. The Aztecs of Mexico spoke Nahuatl. In that language butterflies are *papalotl*. The eggs, which resemble the seeds of the amaranth (*ahauatli*) they called *ahuauhpapalotl.*

What word did our more ancient ancestors, the Indo-Europeans, use for butterflies? Usually, a modern word has cognates in different languages from which we can infer the ancient root. Our English "three" is clearly related to Latin "tres," Greek "treis," German "drei," Welsh "tri," and so on, with the likelihood that the Indo-European root was something near them. But the word butterfly is so distinct in so many languages, it's not possible to find a root. This gives rise to what has been called The Butterfly Problem of linguistics (There's quite another Butterfly Problem in the realm of topology, a kind of higher mathematics, so-called because the problematic figure, inscribed in a circle, resembles butterfly-wings.)

Biologists call moths and butterflies Lepidoptera, for the "scaly wings" they possess. It's not always easy to tell a butterfly from a moth. Moths have feathery antennae while butterflies have golf-clubs for antennae; moths usually hold their wings outstretched when resting while butterflies fold them, though not all individuals of either kind follow that rule. Moths of course are mostly nocturnal, and most butterflies diurnal.

To name something, you must notice it. Butterflies were noticed millennia ago. They show up in some of the earliest Neolithic art which has come down to us, from at least seven thousand years ago, and have made their way into the artwork and folk culture of peoples from around the world. Maraleen Manos-Jones has gathered many cultures' motifs and images in her wonderful book *The Spirit of Butterflies*. She shows them on ancient Minoan frescoes, in Aztec codices, gaily decorating Hopi baskets, and playing their part in Chinese calligraphy.

Moths and butterflies were not only noticed long ago, they were used. As much as we are an aesthetic species, we are also a pragmatic one. The Chinese among others made good use of silkworm moth larvae. To the Muzo Indians of Mexico (see below) monarch fat and protein were a kind of seasonal manna, given to them by the gods. Surely other peoples in other times had their favorite lepidopteran larvae recipes, but most are lost to us.

Sometime after noticing and using comes studying. Surely some early *Homo sapiens – or Homo erectus*, or even before them – marveled to watch a butterfly eclose from its chrysalis, and in whatever rudimentary language he or she possessed tried to pass the word around about where butterflies really do come from. Did they also notice the connection between the caterpillar and the chrysalis? We don't know. Unraveling the details of lepidopteran biology and ecology awaited the rise of science as a way of knowing, something which didn't happen until the last several centuries. Of course that study still continues.

In May of 2007 we lost a pioneer lepidopterist, Charles Lee Remington, who died at the age of 85. For many years a professor at Yale, Remington helped found the Lepidopterist Society and was an avid advocate of the Xerxes Society, which focuses on the study and conservation of invertebrates. Through his love of butterflies he became a good friend of Vladimir Nabokov, one of America's most famous and productive lepidopterists. Lincoln Brower and Robert Michael Pyle were both students of his. His creativity, love of nature and meticulous observation of the natural world, over the years and the eighty or so doctoral students he mentored, have left a legacy deserving mention and respect.

The arrival of Europeans in North America brought many changes to all the land's aboriginal inhabitants. Once the native peoples had been shoved to the side the prairies, which surely had some milkweed plants scattered among them, were quickly plowed and planted into crops – though then, as now, milkweed was likely a persistent "weed." Meanwhile large expanses of forests, not good milkweed habitat, were cut into smaller patches, producing more edges, where milkweed spread and thrived. In the nineteenth-century very large populations of monarchs were noted, perhaps larger than ever before or since. Since most butterflies over-winter in one stage or another rather than migrate, what happened in the winter to these great monarch populations became a question of some interest. There were reports of huge orange flocks streaming southward. In 1885 Pennsylvania naturalist John Hamilton described a stopover site for their migration near Brigantine, New Jersey, and a ribbon of monarchs two and a half miles long by 400 yards wide. Several decades later Roger Tory Peterson, whose *Peterson Field Guides* are so popular, described a huge flock at Cape May, New Jersey. In 1930 C.B. Williams published his book *Migration of Butterflies*. In it he suggested that over 200 species of butterflies migrate significant distances, a hypothesis which was not well accepted at the time and has not held up.

The Russian explorer Otto von Kotzebue was among the first to describe the west coast of North America. Naturalists traveling with him sketched a monarch they'd collected in 1816 near San Francisco. There are published reports dating to 1862 of "a great gathering" of monarchs on Telegraph Hill in that city. The first published account of monarchs over-wintering in large colonies appeared in California's Monterey Weekly Herald on May 30, 1874, when a reporter wrote that "Millions were fluttering around; while overhead stout branches of firs dropped with their weight; and all twinkling in shade and sunlight, seemed the personification of happiness in their leafy home." These coastal California monarch colonies, discussed in more detail below, became something of a tourist attraction. In the early 1930s the town of Pacific Grove passed an ordinance making it a crime to molest the monarchs, subject to a fine of $500. The ordinance may have been enforced against individuals but it was clearly overlooked when developers made their millions while devastating the area's monarch habitat.

Like most scientific discoveries, finding the over-wintering sites of monarchs born east of the Rockies was equal parts hard work, creativity, and happenstance. Falconry is an ancient sport. Trained hawks – and often other birds – were "ringed" with an identification band so they might be returned if lost. Records of banded birds being recovered go back almost five hundred years. Serious scientific banding started right around 1900. Frederick Urquhart, a biologist at the University of Toronto, asked: If birds could be banded, then why not migrating butterflies? In 1937 he became the first to mark monarchs in an attempt to follow their journey southward. At first, Urquhart and his colleagues, including David Waddell, painted marks on monarch wings. By the early 1950s he and his wife Norah had established a method of tagging called alar marking which involves tiny adhesive tags on the monarchs' wings, Latin *ala*. Tags sent back to the Urquharts from many parts of America and Canada suggested a generally southwesterly migration from Ontario. This was useful information. An article was published about Dr. Urquhart's work in the Dec. 31, 1956 issue of Time magazine.[29]

Large migrating waves of monarchs had long been observed in Mexico, but their connection to the monarchs seen flying south in the U.S. had not yet been made. In 1936 Jerszy Redzowski, a Polish botanist who'd emigrated to Mexico, reported seeing a huge wave settle on some mesquite bushes north of Mexico City. In the late 1960s a butterfly tagged by the Urquharts was recovered near the Mexican city of San Luis Potosi, the first clear evidence linking the Mexican and more northerly monarchs. To stir up interest in the monarchs, Norah Urquhart (who passed away March 13, 2009) wrote articles about them for the Mexican papers. In 1972 Ken Brugger, a textile engineer from Wisconsin who'd moved to Mexico after the breakdown of his marriage, happened to read one of the articles. Brugger, who worked for Jockey Shorts and is credited with inventing nonshrinking underwear, was an avid outdoorsman. While hiking in the highlands of central Mexico he came across a migrating wave of monarchs, and notified the Urquharts. In 1974 he married Catalina Aquado, from the state of Michoacan where the monarchs had often been seen. The two of them combed Michoacan on many motorcycle trips, finding more and more monarchs. Finally they convinced a local campesino to take them up into the mountains, where on January 2, 1975 they came upon an over-wintering site with its millions of monarchs.

The peoples of north-central Mexico of course knew about the monarchs. They called them palomas, or "doves." Because they often showed up around the first of November, *el dia de los muertos*, or All Saints Day, they considered the monarchs to be the returning souls of the dead. In the dialect of the local Indians, the Otomi-Mazahuas, they were *seperitos*, "those who come in autumn." In older times these people fried the monarchs in their own fats and feasted on them, fed them by the shovel-full to their cattle, and it is said just for fun sometimes set entire trees on fire, the over-wintering clusters burning like great torches. Chip Taylor (Dplex-l list) notes that indigenous Australians also fed on the fat-rich roasted bodies of bogong moths. Our own ancestors, those of us who are of European extraction, often stayed alive – and increased their fitness! — by being equally resourceful.

Eating monarchs, you ask? But aren't the monarchs poisonous, and didn't the Otomi get sick? Fred Urquhart and Lincoln Brower, the two best-known monarch biologists, had a long dispute over how toxic monarchs really are. Brower was the first to film a blue jay eating a monarch which as a larva had fed on *Asclepias currasavica*, with its high concentration of cardenolides. It took only a few seconds for the poor bluejay to vomit, who then would not eat another monarch. So Brower wouldn't think of eating them. But Urquhart fried some for his friends and colleagues, who said they tasted like toast. Their toxicity – as discussed in a previous chapter – is ultimately a matter of what kind of milkweed the larvae ate, and of course how many and what parts are consumed.

Eventually the Urquharts, none the worse for having eaten the monarchs, traveled to Mexico. No longer young, and suffering from heart trouble,

Frederick hiked up the mountainside with the Bruggers. Here's how he described the end of that journey, in the August 1976 article in National Geographic magazine:

"Then we saw them. Masses of butterflies everywhere! In the quietness of semidormancy, they festooned the tree branches, they enveloped the oyamel trunks, they carpeted the ground in their tremulous legions. Other multitudes — those that now on the verge of spring had begun to feel the immemorial urge to fly north — filled the air with their sun-shot wings, shimmering against the blue mountain sky and drifting across our vision in blizzard flakes of orange and black."

A huge branch of an oyamel crashed to the ground from the weight of the butterflies. The Bruggers and Urquharts sifted through the dead butterflies. They found one tagged in Chaska, Minnesota. There was no longer any doubt where the snow-bird butterflies went for the winter! At the same time the scientists had solved a mystery for the indigenous peoples of the area, who wondered where the monarchs flew off to each spring.

Urquhart would not reveal to Brower the exact location of the over-wintering sites; the two had been feuding as scientists for some years. Brower believed that rather than tagging the butterflies, by analyzing the exact chemical fingerprint of the cardenolides in the over-wintering monarchs, he could identify the region of the U.S. they'd come from. So he set out on his own to find them, and on the day after Christmas of 1976, he finally did. Here's his description of first sighting them on the mountains near Angangueo:

"It was like walking into Chartres Cathedral and seeing light coming through the stained-glass windows...This was the eighth wonder of the world."

In subsequent papers Brower showed that the over-wintering butterflies had cardenolides mostly from Texas. Both his evidence, and the tags found by the Urquharts, proved beyond doubt that these monarchs were gringo tourists from north of the border. Brower has written a nice review article describing the migration of the monarch.[30]

Dr.Chip Taylor runs a very active tagging program out of the University of Kansas. Taylor who has also created the wonderful http://www.monarchwatch.org website, and in media parlance would likely be nominated "Dr. Monarch," for all he knows about them and all the work he's done for them. Each year he closely monitors the southward fall migration from volunteer reports on the Dplex-l list. He uses these reports to predict the overall number of acres the over-wintering monarchs will inhabit in the oyamels. In 2006 he predicted a total of 16 acres, while the actual total came to just under 17. For the 2007-2008 winter Dr. Taylor's prediction of 12.4 acres was close. The monarchs actually settled into 11.4 acres. These habitation areas are early-winter numbers. According to Dr. Taylor (Dplex-l list), the colonies contract by about 10% by midwinter.

More sophisticated analytical techniques have since been applied to the question of the birthplace of the over-wintering monarchs. The result of deuterium and carbon-13 analyses suggests that some monarchs from southeastern Canada and the northeastern U.S. migrate to Cuba. Cardenolide analysis suggests they stay there, interbreed with the resident monarchs, and do not return to the U.S.[31] One study looked at wing shape and angle in migrant vs. resident Cuban monarchs.[32] The wings of migrants were longer and of different wing angle than those of residents. Modern DNA techniques, such as micro-satellite DNA analysis, should further help us learn about the morphology, behavior, history, and movements of monarch populations.

Amazing things are now being done with small digital cameras attached to the backs of birds, allowing us to vicariously experience their incredible flights. Butterflies are far too small for any such cameras. Ornithologists have flown ultralight airplanes alongside migrating geese, cranes, and other birds. In the documentary *Winged Migration* you can, from the comfort of your living room, travel thousands of miles as an albatross, tern or stork. These airplanes have been successfully used to train endangered whooping cranes, raised in Wisconsin, to migrate to Florida. In the fall of 2006 Francisco Gutierrez piloted his hang glider Papalotzin from Montreal to Michoacan, following a stream of monarchs and logging 60-90 miles a day. Though we may not have the skills or time available to join Gutierrez on his flight, it is nice to know that one of us has accompanied them on their long, hard journey to the mountaintops of central Mexico.

Life can be hard. Some times are harder than other times. For us all, birds and bees and coconut trees. One strategy during tough times is to hunker down. Bears do it: they hibernate. Squirrels estivate, which is a kind of partial hibernation. Frogs burrow into the bottom of ponds, sometimes freezing solid, awaiting the spring's first warmth to come back to life. Annual plants toss their seeds to the winds of fate, then die. When it's too hot or dry or cold, trees and other perennials go into a kind of dormancy.

Quite another strategy is simply to get the hell out. It's a decision some of us find ourselves having to make, caught in a bad relationship, apartment or job. Wait it out, or get out? For many animals the decision has been made by evolution – the loud, honking geese, the ducks, thrushes and warblers flying in twittering streams across the face of a harvest moon, and *Danaus plexippus*, one of several kinds of butterflies which leaves when there are no more flowers to nectar, and hard frosts threaten to mow them down.

Many insects migrate, short to long distances.[33] Recent research shows that the common dragonfly known as the green darner migrates along the East Coast of the U.S. to Florida, where they lay eggs, and the next generation moves back north. Some individual dragonflies may migrate 1500 miles or so.[34] That we didn't know about this dragonfly migration until 2008 shows how much we still have to learn about the natural world.

There is a species of milkweed bug on islands in the Baltic which migrates from tens to hundreds of meters from its summer grounds to more protected rocky sites – a migration, however short. A relative, the large milkweed bug of North America, actually follows a similar route as the monarch, though the milkweed bug stops before crossing the Mexican border. So it is not remarkable that monarchs migrate – what is remarkable is the distances they cover, in such great numbers, to so precise a goal.

Most of the monarchs' relatives make their home in the tropics, ancestral home of the danaid family. As mentioned earlier, they appear to have evolved in central America, and came to specialize on the eating of milkweeds. They and the milkweeds go back a long way. It is of the nature of plants, animals and other species, to expand their range where and when possible. At some time milkweeds made their way north through Mexico, into what is now the U.S. Succeeding periods of glaciers drove them south again. While some populations of monarchs remained in Mexico and central America, the more northerly populations likely retreated southward as the cooling Ice Age climate brought winter to them. One way to infer the history of this nascent migration pattern would be more detailed studies of the paleobotany of the south and central U.S. This study, called palynology, uses pollen deposited by local plants over the centuries to construct an ecological and botanical history of the areas.

At first these nascent migrations were likely short. Each fall perhaps a hundred miles southward, to more livable temperatures. But though they might not freeze in central Mexico, winter there is not Panama. There are few flowers to nectar from at that time of the year. Those populations of monarchs which attempted to remain active over winter did not survive. Those whose genes allowed them – urged them – instead to dial down their activity and enter into a kind of estivation did better, and survived to pass their genes on to the coming generations.

Slowly, over the years, they got better at it, this migrating and over-wintering thing. Populations which settled at just the right temperature and humidity, where they were also protected from wind and rain, survived best. These conditions they found in the forests of the high volcanic mountains of central Mexico. It was Goldilocks settling into Baby Bear's room.

To the audience caught up in a spellbinding production of *Hamlet,* or an adept performance of a Bach *Partita,* what we see or hear seems barely possible, verging on the miraculous – a transformation of the mundane material world into a higher, lighter aesthetic sphere. But it is what we don't see that makes this transformation possible — the many hours of practice that came before the performance. So it is with the monarchs. By the time of the retreat of North America's last glaciers some 10,000 years ago, monarchs had been practicing this migration and over-wintering thing for a long time. What we are seeing as we catch a glimpse of one or a wave of thousands riding a northwind south is in fact a virtuoso performance by practiced professionals.

That makes it no less intriguing. How does a seemingly fragile half-gram insect with a brain the size of a pinhead make its way up to two thousand miles to a fifty acre patch of forest it nor any of its traveling companions has never been to before?

The short answer is, we don't know.

But there is a longer answer, and yes you're about to get it. Briefly, the longer answer is, we know a little, here and there, and from these bits and pieces of what we know, we can begin to put together the puzzle, admitting that for now there are still some important pieces missing. But the partial-picture is nonetheless enthralling.

Diapause is a kind of partial hibernation, a pre-programmed shutting down of physiological processes. As mentioned before, in late summer there's no sense mating and laying eggs which, in the cold days of autumn, will not hatch or whose larvae will not have time to develop. So instead of mating most late-summer females enter reproductive diapause.

Reproduction requires energy. Reproductive diapause saves that energy for migration, and shunts those calories instead into fat used in traveling and over-wintering. Hormonal changes occur which make these migratory individuals more tolerant to cold, grant them a ten-fold increase in potential longevity, and somehow turn their thoughts — or urges, if you prefer —

southward. This reproductive diapause lasts until the lengthening days of spring come to the mountains of Mexico.

Modern molecular biology and gene-transfer techniques provide biologists tools they never had before. With these "molecular microscopes" we've discovered and have begun to understand the workings of circadian clocks. These are built-in timers we share with almost all living things, which help us set our physiological clocks to the solar clock. The solar clock is set by the turning of the earth, and the fact that its axis is tilted, as it journeys around the sun. Inside us evolution has composed a rhythm to match the solar clock, a rhythm of genes being turned on and off over the day and over the seasons. The turning on of a gene causes a flush of RNA, which cascades down waterfalls of synthesis into a pool of proteins, which then determine our physiology or behavior. This has all been worked out in remarkable detail in the not-so-distant cousin of the monarchs, the fruit fly *Drosophila*, the insect model for genetics.

Many animals produce proteins called cryptochromes (CRYs). These specialized molecules, which in insects are sensitive to blue light, help set the individual's circadian clock. Fruit flies have one CRY protein, which is light-sensitive and keeps track of daylength. Humans and mice have two, as do monarchs. Human cryptochromes are not found in the rear of the retina, where the cells are found which allow us to see, but instead in the front of the retina, near the pupil. People who are blind from retinal degeneration usually still have functioning circadian clocks. One of the mammalian CRYs resembles the fruit fly light-sensitive protein and the other regulates gene activity, a useful function for a circadian time-keeper.[35] Mutations in the CRY genes interfere with the function of the circadian clock.

Most scientific papers are dour puritans, functional, drab sermons in no-nonsense monotones. Steven Reppert's paper[36] "A Colorful Model of the Circadian Clock" is an exception. It is one of those papers so well-written it makes a difficult subject – the molecular events surrounding insect reproductive diapause — interesting and understandable. Here's what that paper has to say about the ticking of the internal clock in particular cells in the in the insects' brain. In those cells proteins called CLOCK and CYCLE turn on genes called period (per) and timeless (tim). The protein PER, coded by the per gene, inhibits the functioning of the CLOCK and CYCLE proteins. The TIM protein, coded by the timeless gene, stabilizes the PER protein, regulates its transport around the cell, and keeps track, from data sent it by the optic lobes, of the solar clock. The result is a regular ebb and flow of gene products, some of which as they build up in concentration feedback regulate their own synthesis. It is more a complex symphony than a mere mechanical ticking, a remarkable mechanism which sets the insect's physiology to the solar clock, meanwhile changing its setting as the days lengthen or get shorter. To me, the fact that we've come to understand it in so much detail is itself remarkable.

And we're about to understand a lot more about monarch biology and genetics. Steven Reppert's research group is in the process of sequencing the 250-million base pairs of the monarch genome.[37] Reppert's group has already found that some 40 of the butterfly's genes are in some way connected with migration.[38] This is exciting work!

But as we look more closely the remarkable transforms into the near-fantastic. There seem to be four cells in the *pars lateralis* of the insect brain which are the key gears in the clock's mechanism. Four cells. In a sense, these cells are the clock, and their ticking drives the behavior of the monarch in the same way a metronome or the conductor's baton drive an orchestra. These four cells do this by regulating hormone levels, especially that of our old friend juvenile hormone. Decreasing production of juvenile hormone slips the butterfly into reproductive diapause and at the same time grants them longer lives. By interacting with messages brought to their eyes by the optic lobes, these cells serve not only as the solar clocks, but also the insect's time-compensated sun compass, providing both the urge to migrate and a general idea of where to go.

Wait. There's more. In a more recent paper by Merlin, Gegear and Reppert[39] evidence is provided that a time-compensated sun-compass exists in the monarch antennae. Removing the antennae disrupts the ability of migrating monarchs to orient themselves correctly under changing light regimes. Painting the antennae with a black paint also changes their behavior in this regard. So although there is a molecular clock in the brain, it's setting mechanism – like the little dial on the side of analog watches – appears to reside in the antennae. Our model of the monarch's time-compensated solar clock and compass continues to grow more complex, the more we know,

If the monarchs' sun compass was sensitive only to the sun's light in the way our eyes are, they could not orient themselves on cloudy days, when the sun is not visible. To function best, the monarch's sun compass pays attention not to ordinary visual light, but to the polarization of ultraviolet (UV) light, a much more reliable cue.

Our modern lives have mostly removed us from the sky and earth signs our ancestors used to orient themselves in both time and space. With no calendars or fridges to hang them on, ancient peoples looked to nature for information about the seasons. Pliny records how the first rising of the dog-star Sirius as it came up before the sun marked the time of fruit harvest, in mid July, and that Egyptian priests also used Sirius' appearance to predict the rise of the Nile. Others noted that the heat it brought with it ("the dog days") often also brought illness and fever – at that time, likely the peak season for malaria. Phoenicians navigating the Mediterranean used the same star-knowledge as the Polynesians did to find their way across the Pacific. Those of us who hike or hunt the deeper woods know how easy it is to get disoriented on a cloudy day. But sunlight is scattered by the atmosphere in predictable ways, and to insects this polarization plane is visible through clouds. Honeybees flying on cloudy days use it to find their way back to the hive.

The eyes of many species of insects are built in a way that allows them to perceive this plane of polarization. This is accomplished by aligning their visual pigments along the dorsal rim area of their eyes, called ommatidia. Some spiders have special eyes devoted entirely to perceiving this polarized light.[40] These authors have carefully mapped out aspects of polarization perception in the cricket and the locust. Through minute dissection of the anatomy of the eye, following the neurons carrying the eye's information to the brain, and by recordings made within the brain itself, they've learned how these creatures filter out information irrelevant to the polarization compass. It appears that this optical/neural system not only tells the insect which direction they are moving in but at the same time provides them with on-time data about the orientation of their body relative to the polarization plane, should they need to adjust their direction.

Experiments have shown that polarized light also plays a part in monarch navigation.[41] By artificially manipulating the polarization angle of ultraviolet light the navigation of monarchs could be changed in predictable ways. Saumen et al [42] went further. They cloned the cDNAs of the UV, blue and long wavelength-sensitive genes of the monarch. They found that all three are expressed uniformly in the retina. But the rhabdoms in the rim area of the eye, specialized for polarized light, only express the UV gene and not the other two, suggesting it is the rhabdoms that use the polarization angle of ultraviolet light to monitor day-length and direction. These authors also discovered a potential pathway by which the butterfly's circadian clock is connected to those four molecular clock cells set by polarized light.

To test the sun-compass hypothesis Henrik Mouritsen and Barrie Frost, at Queen's University in Kingston, Ontario put monarchs into tiny flight simulators and watched them fly. Flight simulators are used to teach pilots how to fly. They didn't need to teach the monarchs. They all ready knew. What the researchers were interested in was how the monarch navigates and keeps track of time during its trip.

Their flight simulator was an elegant little invention of their own. Individual monarchs were glued to a fine wire with beeswax, set into the apparatus, and lifted gently up with a brush of air. Once airborne, the little guy or girl was constantly monitored by four cameras and an onboard computer connected to optical encoders. The computer continually tracked the monarch's headings for at least an hour of flight.

Fifty-nine adult monarchs were captured on the north shore of Lake Ontario, all nice and fat and ready for the fall migration. These monarchs were kept for five days under differing light/dark conditions. Some were kept in conditions that exactly mimicked the regime of light outdoors. Another group were kept in conditions of day and night with the time moved back 6 hrs, and another group moved ahead 6 hours. If the monarch relied for navigation on an internal clock set to a time-compensated sun compass, the prediction was that the latter two groups would shift their orientation 90 degrees (6 hours being ¼ of the day and 90 degrees being ¼ of 360 degrees) in opposite directions from the normal-time group.

Each monarch was launched headed exactly north. Some of the tethered monarchs were put in artificially cloudy conditions, while others were placed in conditions which had their magnetic fields rotated. These tethered butterflies spent 93% of their time flapping and 7% gliding, their legs like landing gear drawn neatly up to their fuselages. (To see the full text of their article and video clips of the flying monarchs, go to http://www.pnas.org and search on Mouritsen.)

After the experiments were completed, each of the butterflies were untethered and released with a hearty bon-voyage to complete their interrupted migrations.

So what did the fifty-nine helpful monarchs tell Mouritsen and Frost and us about monarch navigation? Control monarchs quickly turned from heading north (snow that way!) to the expected southwesterly direction. Monarchs who'd had their time advanced six hours headed off to the southeast. Those whose internal clocks were running six hours slow instead headed to the northwest, as expected. Butterflies trying to find their way in artificially cloudy conditions got lost, heading off randomly; in these experiments it appears UV polarization was not an adequate guide. Changing the animals' magnetic fields did not affect their behavior. In other words it appears that these monarchs used a time-compensated sun compass as at least their first level of information when deciding which direction to head to (what "vector" the monarchs use to orient themselves, in Chip Taylor's terminology.) Taylor's lab at the University of Kansas, by the way, was the first to indicate a time-compensated sun-compass was involved.[43] It would be interesting to know whether, when, and how the manipulated individuals in Mouritsen and Frost's study, who headed off in the wrong direction, readjusted and changed their flight-path to the correct southwesterly direction.

Other research involving clock-shifting of migrating monarchs captured in Kansas, and then measuring the heading of them after release, supports the inference of sun-compass navigation.[44] Not only monarchs use a time-compensated sun compass to navigate. Similar experiments on other butterfly species captured migrating over Lake Gatun in Panama showed these species used similar navigation aids.[45]

But it is not all sun-compass, for monarchs or other species. Young sea turtles depend on an inner magnetic compass to find the sea after hatching. Researchers tethered 25 sea turtles in a backyard pool with a strong magnetic field around it. Released, they headed off towards home – Melbourne, Florida – in the direction their magnetic compass told them to go. Switching the magnetic poles 180 degrees caused them to head 180 degrees away from Melbourne. It also appears the turtles imprint a magnetic location on their birth-sites, so they can later return to it.[46] Recent evidence suggests migrating birds use their eyes to "see" earth's magnetic field.[47]

Redundancy on a piece of paper is not good. Is not good. But when the loss of a critical function is potentially fatal, redundancy takes on real value. So we build it into the space shuttle's many systems and into the controls which fly commercial aircraft. Evolution has built several levels of redundancy into monarch navigational controls. In addition to a time-compensated sun compass the butterflies seem also to carry a magnetic compass. Captured fall-migrating monarchs placed in an enclosure without any magnetic field fly off in all directions. If the artificial magnetic field mimicked the natural field, they took off to the southwest, the correct direction. But in a reversed magnetic field, in which south appears on a compass as north, most headed to the northeast.[48] So in addition to their sun-compass, they do make use of a magnetic compass. In nature both compasses will give them the same instructions, which must in some subtle insect-y sort of way be reassuring.

Animals which possess a magnetic sense must carry some structure with magnetic properties. Preserved monarchs were magnetized and dissected. While in bees most magnet-sensitive materials are found in the abdomen, the monarchs' compass seems localized in the head and thorax. Their magnetic material was able to register the direction of applied magnetic fields, a property useful for navigation.[49]

Experiments in the late 1990s showed that butterflies migrating in Panama used a sun compass. More recently individuals of the species *Aphrissa statira* were captured while migrating over Panama's Lake Gatun. Control individuals were kept in a magnetic field which mimicked the one outside while others were exposed to a magnetic field shifted 180 degrees. Most of these headed off in the opposite direction they'd been flying in. The direction the monarchs headed depended on whether they were allowed to use their sun compass or not. If they were they headed, correctly, in a direction relative to where the sun had been when they'd been captured. If they weren't allowed to use their sun-compass, they flew off either in the correct direction or 180 degrees off, suggesting they use both the magnetic and the sun compass to navigate. When the data from both compasses agrees, the butterflies fly correctly. When the compasses have been artificially manipulated to disagree, one or the other compass takes precedence.[50]

Each of the sciences has its honor roll of classical experiments – Priestley's discovery of oxygen, Thomas Young's double-slit experiment demonstrating electron interference, the response of Pavlov's dog to a lunch bell. Niko Tinbergen was one of the first ethologists, biologists who study animal behavior. It was he first pointed out that hatchling birds imprint on the first live creature they set their eyes on – whether momma bird or big lanky biologist like himself. Tinbergen also showed how the digger wasp finds its way back to a nest it's just dug. He placed a circle of pine cones centered around her nest, and waited for her to leave. While she was gone, he moved the cones a short distance away, and when she returned she headed not for the nest but for the center of the circle of cones.

T.S. Collett [51] observed that Heliconiad butterflies from Central America somehow find the same branch to roost on every night. Collett suggested that when near their goal, insects compare the landscape to a remembered retinal image of it, a kind of snapshot stored away in their brain. Collet showed that if you train them that food is a certain distance from an upright cylinder, then increase the size of the cylinder, they land far enough away from it that the cylinder matches the size of the retinal image they had remembered. Experiments with honeybees suggest they use a combination of distant landmarks to locate themselves towards a goal, then switch to more nearby, remembered landmarks which they compare with this remembered retinal image.

Bees and wasps often perform orientation flights after eclosing, to learn the landmarks of where they were born. Males who fly on a one-way trip to mate and then die do not perform these orientations, but females who expect to return do. One study indicated that a solitary wasp often did an orientation flight every morning. After crawling out of her nest – a hole in the ground – she would turn to face it, then fly increasingly large arcs around it, all the while keeping an eye on it. Then she would turn and fly around the nest while watching it with the opposite eye, in that way remembering the nest in relation to landmarks visible with both of her eyes. When she returned, in order to match the remembered image to the one she actually saw, she usually came in on a flight-path which coincided with one of her orientation arcs. Experiments indicate that this up-close landmark matching only works within a few meters of the nest: the insects must use another mechanism to get that close.

When farther from home the landmark's distinctive shape is remembered, not so much its size. Landmarks are used as a directional guide to go straight or turn at a certain point. Once a particular landmark is behind the insect, it guides itself by the next one in its memory. This way the entire "map" does not need to be memorized, just one page at a time, along with the sequence of "pages." Foraging bees are able to store the maps for a series of flowers, one after another. Some tropical bees are known to forage at and remember about 40 different isolated plants on a 20 km path of tropical forest.

As Robert Pyle shows in his travelogue of more than 9000 miles following monarchs from Canada to California (*Chasing Monarchs*), there's much more to the monarch's navigational skills than just a time-compensated sun compass, tweaked by the magnetic field. Monarchs, like bees and like us, use landscape cues —mountains, valleys, and especially river valleys — to find their way. Other research corroborates the importance of river valleys.[52] A distinct advantage to flying up a river valley is its lack of abrupt changes in elevation. Another is that it leads you somewhere predictable. Human pilots know this. They've been following river valleys since we first left the ground. The terrorists who flew into the Twin Towers flew up the Hudson River to their final, fatal destination.

Some studies, controversial because they've not been readily replicated, suggest that homing pigeons may use their sense of smell to help find their way home. Orientating by smell is common in fish; after several years in the open ocean, salmon find the stream they were hatched in by its distinct smell. It seems possible to me – though I don't know this has been tested – that a genetic memory of the odor of oyamel firs has been encoded into the monarchs. One theory has it that once nearby they find the ancestral over-wintering sites by smelling the dead bodies of their ancestors, decomposed under the oyamels.

So the evidence suggests that in open terrain migrating monarchs use either a compensated sun compass and/or their magnetic compass to navigate, often finding advantage in flying river valleys. They are likely to take different routes, depending on the weather and other as yet unknown factors. In other words, their flyways over the south-central U.S. are not exact, fixed routes. Migration begins as early as mid-August for the most northerly monarchs. If you're wondering whether the main southward migration has passed you by already, from tagging data Chip Taylor (Dplex-l list) suggests that monarchs who leave a particular latitude after the angle of the sun at noon is less than 47 degrees aren't likely to make it to Mexico. For a latitude about that of Minneapolis-St. Paul (around 45 degrees north), that means that because of cooler temperatures, fewer milkweed to nectar on, etc., migrating monarchs need to leave before about Sept. 15. The sun angle at the over-wintering site when most migrants arrive is about 57 degrees. They leave in the spring when it reaches approximately that same height in the sky – sometime after Valentine's Day.

Fred Urquhart was creative in using the tagging method to learn about monarch migration. He asked himself: Where would monarchs from the Toronto area migrate to if you caught and released them in British Columbia? In the 1960s and 70s some 900 Ontario butterflies were tagged and shipped to Vancouver to be released there (Don Davis, Dplex-l list.) Those that were recovered were found in B.C., Washington, Oregon, and California. This tells us they don't have a destination somehow burned into their brain, towards which they fly. Instead, these monarchs joined their buddies and got on a flight to California.

Once having arrived at Mexico's Sierra Madre Oriental eastern monarchs likely use visual cues—though these cannot be remembered cues, since they haven't been there before — to follow the mountain ranges to their over-wintering sites.[53] It turns out we know quite a lot about how the monarch migrates, and it's a complex story, but we still have lots to learn.

The genetic encoding of complex navigational information in birds was demonstrated when individuals of a migrating species of European warbler were captured and allowed to "migrate" in special orientation cages. These cages allowed the scientists to determine what these birds' preferred direction of flight was. Though not actually covering any distance, the birds tried to head off in a particular direction for several days, then changed their heading to fly around mountain ranges they would actually be encountering were they flying. These were young birds who'd never actually flown that route![54] Hypothetically monarchs could come equipped with a similar kind of map, which they refer to as they approach the Mexican highlands. The fact that we have no idea what format that map would be written in makes this another of the many intriguing mysteries facing new generations of young naturalists.

A recent study suggests that pigeons use a combination of compass direction and memorized landmarks to find their way around familiar territory.[55] In other words while migrating birds and butterflies have available to them a menu of navigation aids. These include sun and magnetic compasses, navigating by the stars — not used by butterflies, who migrate only during the day —and landmarks in the terrain, which may or may not be compared to an innate "map." These various mechanisms are self-correcting, each calibrating the other. Hand-raised savannah sparrows are born with a magnetic compass. But by exposing young sparrows to a kind of artificially-rotated star-map, the birds adjust their compass to fit the stars.[56] Which navigational aid the birds or monarchs use at a particular moment may depend on a variety of factors, including their own assessment of the reliability of that aid based on recent experience. Predicting monarch behavior, like predicting human behavior, is no walk in the park.

In addition to all these navigational aids butterflies seem to come equipped with plenty of good common sense. Schmidt-Koenig[57] took vanishing bearings of 1,599 monarchs headed south, from Long Island and Ithaca New York south to New Jersey, Baltimore and North Carolina. Almost all individuals flew in a south-southwesterly direction. Those who deviated from this did so because of heavy winds, or to fly along the lee of woods, dams and large buildings. They preferred not to head out over the ocean. If there were strong offshore winds, they would take off, trying to fly southwest, but if the winds carried them toward the ocean, they would turn around, flying low over the water to return to land, where they would wait a while and try again. This may explain how large migration clusters build up, waiting for good winds. They made good use of tailwinds, and were seen to fly equally correctly in sunlight or clouds, sometimes at heights only visible through good binoculars, flying against high clouds with kettling hawks. Monarchs have been seen slowing down over rising thermals, such as from large parking lots, to circle and gain altitude with the rising air-currents. Broad-winged hawks have also been seen lunching on migrating monarchs kettling alongside them. Three species of migrating butterflies (cloudless sulphurs, gulf fritillaries, and the long-tailed skipper) were seen to fly low to the ground on especially windy days on their way to Florida. By staying low to the ground they flew in the boundary layer which experiences less wind or turbulence, a trick also used by pelicans, cormorants and other seabirds.[58] Low tag-recapture rates of monarchs flying along the coast, compared to those migrating inland, suggests that coastal routes may be the more dangerous. In fact swaths of butterfly bodies sometimes show up along the shore. During daylight hours breezes tend to be land-ward, and would provide some lift to coastal-migrating individuals. But after sunset the early night winds become off-shore winds, decreasing lift and possibly carrying coastal-migrating individuals out to sea, where they drown and are then washed ashore.[59] A natural question then comes: Do fish eat floating dead monarchs? One answer, supplied by Jim Miller on the Dplex-l list, is that while the fish in his aquarium eat many kinds of dead butterflies, they've never accepted a monarch, even on first try. Is this because the monarchs release some of their bitter defensive compounds as they float into the water, making them unpalatable to the fish, or are the fish born with a genetic knowledge not to consume them? These are testable hypotheses.

Carol Cullar[60] reports watching several monarchs flying, downwind, past a growing cluster of migrating monarchs. Just as they pass the cluster, they turn and flap back upwind to it, as though they have intersected and detected an odor plume from the cluster. That, in addition to seeing the cluster, may be how they find one another. She and others also report that towards dusk monarchs heading out over the Gulf of Mexico, having lost sight of the land, will turn around, fly north to the shore and settle in for the night before heading out in the morning, as though realizing that this is not a small lake but instead a wide body of water over which they could easily get lost during the night. Another interesting observation during the southward migration features the Golden Orb Weaver, a slightly poisonous spider common in Texas. This species often lives socially, sometimes dozens building their webs in one tree or shrub. In the fall, just as monarchs are heading past, young orb weavers disperse to new sites by climbing to a high branch and sending a short string of web out into the wind, which lifts them into the air. Sometimes the string attaches them to passing monarchs who provide them a free ride to a new dispersal site. Carol also reports (Dplex-1 list) that monarchs aren't the only butterflies to cluster. She has also seen common zebras (*Heliconius charithonius*) cluster in cold weather.

There is good evidence of a monarch migration route from Louisiana across the gulf to Mexico. Migrating birds often land on oil rigs off the coast. The rigs, often painted a bright yellow, also seem to attract migrating monarchs who are either fooled into thinking the rigs are patches of wildflowers or are simply using them as a place to rest. They appear to be able to locate the rigs in the vastness of the Gulf-waters. Whether they do that by sight or perhaps sensitive magnetic navigation remains unknown.[61]

My ninth edition (1870's) *Encyclopedia Britannica* has a long article on navigation, redolent with diagrams of astrolabes, chronometers and trigonometric formulae. The monarch's navigation system is immeasurably better than the one Columbus used to stumble onto the shores of the New World, and works just fine in a brain hardly bigger than a grain of sand.

One important aspect of navigation is to head off in the right direction. Another is keeping track of how far you've gone, so you don't fly right over wherever you mean to go. Some insects are equipped with a kind of odometer as standard equipment. Waggle-dancing bees communicate the direction of a good nectar source to their hive-mates by waggling their butt a certain number of degrees from vertical. They communicate the distance to the nectar by waggling for a shorter or longer time. Obviously to communicate that distance, they must know it. While foraging they've kept track of how far they've had to fly to various nectar sources by recording in their brain how much visual flow they've perceived, which is a kind of bar-code for distance. In an object-poor landscape, such as a flat featureless field, a certain duration of waggle stands for a further actual distance than the same amount of waggling done in a visually-rich landscape: in this way, the dance is made relevant to the local environment. One estimate is that a millisecond of bee-butt waggle corresponds to 17.7 degrees of image-motion within the eye.[62]

Foraging ants are paradigms of selflessness. Like the Russian agronomists in World War II who were told to care for a warehouse full of seed potatoes and starved to death rather than eat them, ants regularly lift and carry food back to the nest rather than eat it. If the food item is too big to handle by themselves, they don't keep it as a secret stash to return to when hungry. Instead they hurry back to the nest and excitedly tell their nest-mates about it, and the nest-mates swarm out to help drag it home. The odometer of foraging *Cataglyphis* ants is not just a pedometer which clicks off how many steps the ant has taken away from its nest, but is also capable of doing some pretty fancy math, depending on how much topography the ants encountered, by converting that pedometer measurement to an actual horizontal distance. Because they may take slightly different paths over rough terrain (the world is much grainier to an ant than it is to us), before they head out the nest-mates must know the distance to the food "as a crow flies." They use both of these separate odometers, one for the number of steps and the other for actual horizontal distance to pinpoint where to find the food they've been told about. We don't know if mechanisms like these are used by migrating monarchs. But because they've never been to the over-wintering sites, they have no memory of the distance they need to cover. Instead, it's likely they find other monarchs, by sight or smell, at the sites, and settle with them.

While we still have much to learn about the details of the magnetic and time-compensated sun compasses of insects, their complexity and the elegance of their mechanisms are remarkable. It's like peering inside a fine Swiss watch, built on the scale of nanotechnology, achieved by millions of years of evolution. To me its equally impressive that our brains, regulated as they are by their own circadian clocks, are able to dissect and understand this mechanism. Learning about the monarch's genes and hormones will in some small but possibly important way inform us about our own genes and hormones. If a healthy long life is a social good, and these migrating and over-wintering orange friends can lengthen theirs by a factor of five or more, there's reason more than elegance and curiosity to put our ears up to the ticking of their internal clocks.

The long warm days of summer are over. The first sumac and red maple leaves blush in a motif soon to be fully developed into a symphony of color. The days are perceptibly shorter, the nights cooler. Hormonal changes took place while I slept in my chrysalis, and on eclosing instead of mating and multiplying I'm destined to flap my way through two thousand miles of obstacles to some Mexican hide-out. What do I need to get me there?

Navigation system?

Check.

This thing I propose to do is not an easy thing for a half-gram insect. It's going to take a lot of will-power, focus, intention, audacity – call it what you want. You don't fly that far against those odds on a whim, or as a wimp.

Watching a butterfly as it flits through a meadow or field, searching for nectar or just the right leaf for an egg is itself a kind of good poetry, carrying us to unexpected little revelations, intent well hidden. But the same butterfly on a cool September day riding a north wind almost shouts intention. Fully focused; no multi-tasking allowed. Stand aside, she says, I'm going somewhere. If poetry, it's not lyrical but epical. Odysseus' homeward journey.

I must admit it is likely no more than another of our illusions, this assigning of purposeful clarity to the animal world, only another way our own emotional needs transform the real world. Having discovered, to our horror, the rot at the core of our hearts, we feel the need to hand off the illusion of clarity to someone else. While feeding the chickadee certainly pays attention to its stomach, the lay of its feathers, the sky and ground and the dangers they might hold. The migrating monarch may simultaneously be on the lookout for nectar, a resting-perch, or a cluster of conspecifics to spend the night with. Like us, her heart drives her simultaneously in different directions, and it is the play of that heart on the instrument of opportunity which constitutes the music of her life.

I also admit using the language of will-power may be slipping over into the cardinal sin of ethology, anthropomorphising. In that tiny brain, in the complex circuitry of its neurons, is there really a sense of purpose, of mindfulness towards a far-off goal? Hunkered down on a drizzly cheerless day, is she frustrated, impatient? Does an uncooperative wind heading her off course bring to her a tiny wisp of anger? Does she rejoice while flowing into a growing stream of her kin, or feel lonely if separated? Having arrived on the oyamels and settled in with her fellow-travelers, does she sigh and feel a sense of accomplishment akin to that we feel when we finally collapse at home or in the motel room after a long trip?

Is there a mind in there? The study is called "theory of consciousness," and as you might expect there's more than one take on it. Some would allow her only a kind of robotic intelligence, the neurons firing like a well-timed motorcycle engine, the hard-wired circuitry taking her blindly southward on automatic pilot, flying by wire.

Others, myself included, caution not only about anthropomorphising, but about under-estimating other critters' minds. Almost every week we learn more about how we have underestimated the intellectual and emotional abilities of dolphins, ravens, elephants. So tell me she's nothing more than a robot. I'll listen, quietly, poker-faced, then play a trump-card. Or two. One is based, like all good science, on pure observation. Watch them. They exude purpose. Bring a monarch caught while migrating into a cage and watch him or her beat her wings time and again against the bars, trying to escape. If but a machine, then I say a machine not unlike that of our own brains and their emergent minds.

That's the second trump-card, the one called "emergent properties." Walk along the shelves of an auto-parts store and pull a starter off the shelves, grab a differential, four wheels each with their brakes, a chassis and steering-wheel, a complete engine if you can get someone to help you heft it, and so on. Pile them out in the parking lot. No evidence, to the uninitiated, that that particular pile of plastic, wires and metal could carry you to San Francisco. But put them together in the right way, and it's time to pack your bags. The property of being able to move and while moving carry a load of people and their stuff emerges unforeseen out of the collection of parts. It's a characteristic of living things, too. Liver, spleen, capillaries and lungs do not make a person. Put 'em together right, and there you are, or me. Natural selection has been playing that particular card for thousands of millions of years.

Maybe, just maybe, from that miniscule brain, its ions and neurotransmitters fluttering in their own waves and streams toward the postsynaptic junction, maybe from that tiny but intricate sculpture of cells do emerge simple insect thoughts, felt desires, and maybe even a flash of joy now and then like that which lights up our own bigger, and yes more complex brain. I have to say maybe, because until someone actually finds that hunger, joy and loneliness in there, it's mere hypothesis, built on a fairly weak inference. Properties emerge. Emotions are emergent properties. We don't know what the minimal brain size or complexity is in which they can emerge. We do know though that we've been wrong about that in the past.

Navigation system.
Check.
Will power — or something like it.
Check.

What else do we need before heading south?
Well, no weak inference here.
Energy!

The job of a monarch which has eclosed early enough in the summer is to reproduce — make more of your own kind. To do this, you sip nectar and shunt as many as possible of your calories and available protein into either your eggs (if you're of the female persuasion) or your sperm and nuptial gift if you're male.

But if you're a late-bloomer, and there isn't time in your northerly latitudes to raise a whole new generation of young, you're asked to put off the pleasures of reproduction in favor of a very long flight south, followed by months of dangerous over-wintering. If you survive those, then when spring finally comes, like a swimmer who's held their breath for a long underwater swim, you hit the surface, inhale, and mate. The cool fall nights and short days have prepared you for this by adjusting your physiology. For the last two weeks or so every calorie you eat goes into fats to fuel your flight and over-wintering.

Ethanol is a biofuel much touted as a replacement for petroleum-based gasoline. But ethanol has considerably fewer calories per gallon than gasoline. That means a car running on ethanol – or E85 – gets fewer mpg than one running on straight gasoline. So, to go the same distance between refills you need a bigger tank.

Fat is the high-energy fuel of choice for hibernating or migrating animals. On a gram per gram basis it's twice as energy-rich as protein or carbohydrates. Hibernating bears build up a nice layer of fat which they burn slowly snoring away in their dens. Migrating birds can put on half their weight in fat before heading off on the long trip. Monarchs do the same. Mobilizing fat and using it to power wing muscles is somewhat more metabolically complicated than using carbohydrates such as the insect sugar trehalose, or the amino acid proline, or ketone bodies. So the monarch relies on these latter fuels for short bursts of speed – like kicking in a turbocharger, while for long, sustained flight such as in migration, it reaches more deeply into its fat supplies. The metabolic mechanisms which "decide" whether to use anaerobic (carbohydrates) or aerobic (fats) to power our monarch are a fascinating subject of their own, but beyond the scope of this book.

Few find the study of metabolism and the ways the body adjusts it to meet short and long-term goals interesting enough to check a biochemistry or metabolism textbook out of the library. I'll respect that metabo-phobia by keeping this discussion of monarch fat metabolism nice and lean.

In our bodies fats are stored in adipose tissue and the liver. Insects store their fats in special tissues called fat bodies, almost entirely as a molecule called triacylglycerol.

As our monarch wings her way southward (or, as she sleeps the winter away) and her energy levels begin to decline, a hormone called adipokinetic hormone is produced by her corpora cardiaca, situated just rear of the brain. As in our bodies, one hormone often regulates the levels of another. In this case, production of adipokinetic hormone is increased in the presence of another hormone, a protein called octopamine. The flush of adipokinetic hormone causes fat bodies to release fats into the blood, called hemolymph in insects. This begins a steady drip-drip of fuel to keep the insect's engines running. Once in the hemolymph the fat is carried by special proteins called lipophorins which are not unlike the specialized lipoproteins that transport cholesterol in our blood.

For the migrating monarch the important thing is to keep those flight muscles moving. Because they can flap and glide, their energy demands are not nearly so high as those of a flying bumblebee or locust, whose metabolic fuel consumption is perhaps the highest of any known animal tissue.[63]

How much fuel are you going to need to fly to Mexico? It depends on several things, really. One is where you're starting from – some monarchs are a few hundreds of miles from their over-wintering sites, others more than two thousand. Another variable is how many refuelings you can manage on the trip. You don't have to fly nonstop. Through analysis of cardenolide concentrations in over-wintering adults, it appears that *Asclepius syriaca*, a southerly species of milkweed, is the main southward food source.[64]

Monarchs are opportunists, and nectar on many kinds of flowers during both the summer and during migration. One spectacular congregation of monarchs was seen in late September of 2007 when more than a hundred thousand stopped to visit a large field of late-blooming sunflowers (Dplex-l list.) But like finding gas stations in the desert west, nectaring stations along the way are not all that predictable. A monarch departing from Minnesota would have no way of knowing about a regional drought or early frost in Kansas or Texas. And don't forget that you need to arrive at your mountaintop retreat with enough fat to get your through the long winter. One study suggests that nectaring along the way is not done to provide fuel for the trip, but to refill the fat stores for the long winter.[65] So, you'll find some nectar along the way but because you don't know how much, it's best to start out with as full a tank as you can. Of course, an important variable is how energy efficient you are, how many calories of fat you burn for every mile flown.

The following table compares the miles per gallon, as it were, of a monarch in normal flight – neither up against a constant headwind nor gliding down a tail wind, with several other modes of transportation.

Monarch	Walk*	Bicycle**	Prius***	747 ****
calories per mile 6	88,000	31,000	481,500	1.7×10^8
cal/mile/gram total wt. 12	1.25	0.39	0.43	0.43
cal/mile/gram passenger 12	1.25	0.44	8.9	6.0

* Walk – assume 4 mph, 155-lb. person
** Bicycle— assume 155-lb. passenger, 20 lb. bike, 15 mph
*** Toyota Prius— assume one 155-lb. passenger (the driver), 50 mpg.
****747— assume 747-400; 416 passengers at 150 lbs each, maximum take-off weight 875,000 lbs. Cargo not considered. Cruising efficiency of 5 gallons/mile used in calculations; take-off and landing are much less efficient.

The calorie used above is the "small calorie," one thousandth of the Calorie usually used in human nutrition, which is actually a kilocalorie. Total weight is vehicle plus passenger. (Note: these are approximate calculations. Energy efficiency values reported vary considerably.)

To estimate monarch flight energetics, I used a value of about 1 calorie per gram per minute, determined for flight of the cecropia moth *Hyalophora cecropi*.[66] Because the cecropia moth is considerably larger and heavier than a monarch, this number might over-estimate monarch energy use. Monarch flight speeds during migration are not well-documented. For tethered flight, Altizer and Bradley[67] estimate about 3 mph, though it seems to me the apparatus may have slowed the butterflies some. Monarchwatch.org suggests a migratory speed of more than 10 mph may be likely. I assumed 10 mph, and a 0.5 gram butterfly weight.

These calculations indicate that as delicate and poetic the flight of a monarch butterfly might be, it's not all that energy efficient. It's a lot of work! And remember that a monarch who arrives at the over-wintering site with an empty tank, or even half a tank of fat, is essentially doomed.

Navigating a gauntlet of dangers as they are, it is not unusual for migrating monarchs to lose part of a wing or many of their protective scales while on this perilous journey. Any such thing which makes flying more difficult will, of course, increase their demands for energy, and decrease their chances of either getting to the over-wintering sites, or arriving there with adequate lipids.

If our migrating monarch has managed to load up until it is about 50% fat, that's about 250 mg of fat. If the fat is worth about 9 calories per mg, that's about 2250 calories of fuel onboard, which if my calculations are correct could carry the monarch about 750 miles without any refueling, at best. If there are headwinds, or cooler temperatures which require burning fat to warm, its nonstop range could be considerably less than that.

That means on the way south there better be some service stations. You can help them on their way by creating a waystation in your yard or garden for them to stop at. Learn more at http://www.monarchwatch.org/waystations. It's important, after all, that they arrive at the over-wintering site with a full tank.

While on the subject of the weightiness of a butterfly, here's a rough translation of a fable by the 18th-century Polish poet Ignacy Krasicki:

The Wagon-Driver and the Butterfly

A wagon got stuck in the mud, and couldn't move.
The driver was tired, the horse exhausted.
A butterfly, sitting on the wagon, thought:
    "I'm told compassion is a virtue."
He flew off, calling back to the wagon-driver:
    "Now get going, bless you!"

Krasicki seems almost to have stumbled upon, two centuries ago, what is known these days as the Butterfly Effect. In nonlinear dynamics, tiny effects can have great consequences – a butterfly flapping its wings in Montana, according to this idea, can affect the weather in Manhattan. It's part of chaos theory. Well, I do hope they got the wagon unstuck...

OK. Ready to roll?
Navigation system, will power, energy, all check.
Whoops.
Almost forgot. We do need a vehicle. Not just any will do – not a child's first tricycle, not a kayak, not even a Lamborghini Diablo. No, we need one that can fly. For long distances. And weighs well under a gram.

As every new pilot soon learns, once you lift your feet from the ground the air most of the time is a very turbulent medium. That turbulence becomes more prominent the smaller you get. That an eagle can fly long distances through and sometimes even use that turbulence is remarkable. That a half-gram butterfly can do the same is but a whit shy of miraculous.

A quick preflight check of the aircraft: Wings to provide lift and thrust, aerodynamics that reduce drag, and an airfoil shape which effectively reduces wing-tip drag.[68] (Long-distance flyers like albatrosses often have long, narrow wings.) The most effective wing and airfoil shapes are different for powered flight vs. gliding. The requirements for gliding are one thing. Even clunky flying squirrels and flying fish can glide. But butterflies, like hawks and eagles, can also soar, that is gain altitude by using updrafts while gliding.

During active flight a monarch's wings do not move straight up and down. While they're moving down they also move backward, to provide thrust. As the downward stroke ends, the wings move upward at a somewhat different angle. In other words, if you could mount a tiny light on the tips of the butterfly's wings, you would see an asymmetrical ellipse being made in the air, not a simple up and down motion. The downstroke takes longer than the upstroke, too. If you've ever spent much time watching a butterfly in flight, you'll also know that it's not a simple repeated pattern. Unlike the steady beat of a gull's wings as it heads home along a darkening coast, the butterfly's wing-beats are very irregular, as is its flight-path. This poetry-in-motion has partly to do with their light weight, making them more susceptible to slight changes in the air currents, and partly to do with the large wing surfaces they use to generate flight. All this battering about by the winds of change of course requires countervailing movement by the monarch to maintain its heading; and that, too is part of the poetry, and the astonishment that such a tiny brain can bring it off.

Turning can take place in several ways: flapping one wing a bit more than the other or raising one slightly while in a glide. Grasshoppers have been observed bending their abdomen towards a turn, which shifts their weight and changes drag. I suspect butterflies do this, too. In many insects the brain compares the visual flow past one eye with that flowing past the other. Less flow out of the right eye, e.g., suggests the insect is actually turning in that direction; it responds by turning in the other direction.[22] But the exoskeleton of insects like butterflies is also covered with wind-sensitive hairs, which along with their very sensitive antennae they likely use as guides to the direction and speed of their flight.

Insects were the first to evolve flight. Because it requires lots of energy, and aerobic metabolism is the most efficient kind, this could only occur when the oxygen in the atmosphere increased significantly some 300 million years ago. The wings of insects are built entirely differently from birds' wings. The thin membrane of the monarchs' wing is made of scales of chitin, arranged in layers. The scales of chitin are roughly a millionth of an inch thick, a thickness which is not uniform. The approximate thickness and roughness of the scales, according to some physicists, makes the scales much more effective at absorbing the sun's energy, and provides the wings their lovely iridescence. Their structure has been used by engineers to fine-tune the efficiencies of computer chips, flat-panel television displays and solar panels.[69] The individual scales also allow a canvas on which the pattern of colors in each species' wings can be built.

But the butterfly wing, held together by a strengthening framework of veins, is not as controllable as the bird's wing, in which muscles can change the warp and woof of the feathers. A fly's wings are stiffer than the monarch's, giving the fly more control over its flight. The broad butterfly blade can only partially resist bending and twisting, which make its flight more subject to the slings and arrows of outrageous air currents. The wings, in a sense, have a mind of their own, which a monarch must learn to work with rather than against.

Insects may not have gravity sensors, and seem to rely instead on vision to tell them up from down; they keep the lightest part of their visual field, the sky, up. Experiments suggest that in total darkness many can't make that distinction. Dragonflies use their head, the biggest and heaviest part of their body, as a kind of pendulum, to compare against what's happening to the rest of their body. If a gust of wind twists their body one way, their head will tend to stay straight. Hairs on their neck detect the twist of the body, and they adjust their flight accordingly. True flies – dipterans to biologists – use the remnants of one pair of wings, called *halteres*, as gyroscopes, which they vibrate while flying. Any variance from straight flight bends the *halteres*, which contain nerves informing the "pilot" they're no longer on-course. Flies, unlike other insects, can fly well in total darkness.

Stand alongside the narrow curvy roads of southern France as the peloton whizzes past downhill. Some riders may be pedaling, while others coast. Watch a flock of monarchs going past in migration and you'll see the same behavior. Gliding is coasting in the air; it rests the muscles momentarily and gets them ready for the next uphill climb. It seems there's always another uphill climb.

As mentioned above, because their toxins provide them protection from predators, monarchs are not the aerobatic flyers some other butterflies are. If you've watched a pair of male red admirals in *mano a mano* air combat, the difference is clear.

Oh, right, and we do need an engine of some kind to move the wings.

That would be the muscles. Insect muscles like our own are made of some of the commonest proteins on earth, actin and myosin, arranged into what are called sarcomeres. The sliding of one protein fiber across another causes the muscle to contract. The energy to do this is supplied by molecules of ATP, adenosine triphosphate, which is the energy source for almost everything that keeps you alive – thinking, seeing, digesting, moving ions, and so on.

Muscle cells are about 25% efficient. That's not bad, for a motor. About the same as a properly-tuned gasoline engine. Diesels are a bit more efficient. Though diesels powered the Hindenburg and other lighter-than aircraft, they haven't yet caught on for propeller airplanes. As aviation fuel becomes more and more expensive, they may power the private plane of the future (see http://www.dieselair.com). Jet fuel is essentially diesel fuel; over time jet engines have become more efficient and at cruise are now in the 40% efficiency range. Electric motors are very efficient, offering back up to 90% of the energy delivered to them. So 25% efficiency isn't bad, but it does mean that 75% of the energy in each ATP molecule used to power flight gets wasted as heat. That's the heat that warms us when we run or shiver, and warms a bumblebee or butterfly on a chilly morning. But it's also the heat that can build up in the thoracic muscles of a monarch flying on a hot day, and urge him or her to take a break in the shade.

And so after eclosing they launch themselves, one at a time, to head off and join the growing stream of southward migrants. Having raised and launched them myself and with my students, it is remarkable to watch the assumption gown carrying them upward for the very first time. In that instant they have a lot to learn, with a very steep learning curve. In my experience they fly a few moments, often doing an orientation spiral, before landing in a tree or on the side of a building. Catching their breath? Resting their new, untried muscles? Overcome by the magnitude of what they'd just done? Opening themselves to a new sense of freedom and adventure?

Meitner et al [70] had the help of the Monarch Butterfly Project in counting more than 22,000 monarchs over seven years at a fall migration stopping-over point at Peninsula Point, Michigan. Some years they counted almost ten times as many as other years. There seemed to be no effect on the numbers in the years following a heavy die-off at the over-wintering sites. It was estimated that three-quarters or more of the over-wintering monarchs died in the winter of 2001-2002. One study [71] of summer populations in Duluth, MN showed a significant downturn in monarch density for subsequent years, but total numbers moving south from across the country in the fall of 2002 were not significantly low. This suggests that a good summer's reproduction can bounce the populations back from near-disasters in Mexico. Cooler temperatures and cloudy weather decreased the counts of monarchs which stopped at Peninsula Point, probably because fewer were migrating under those conditions. The butterflies stop to rest at Peninsula Point and to replenish their fat stores before moving on again. In one tagging study in Virginia, the average stay-over was two days.[72] In the fall in west Texas a breeding generation of monarchs can be found in mid-September, and eggs are sometimes found as late as mid-November. These could be local residents, or reproductive early migrants from farther north. The main mass of migrants coming from the north arrives in October. In 2007 it peaked in mid to late October, with reports of large clusters and unusually large flocks seen headed south, which some described as "rivers of monarchs." The main fall migratory pathway is west of Austin[73] though there is another pathway that follows the Gulf of Mexico (see this website[74] for a nice map.) One suggestion is that the more western pathway is made up mostly of Midwestern monarchs, the coastal pathway of eastern individuals.[75] Another more coastal migratory pathway takes monarchs from the east and northeast down to Florida, where some apparently hop across the Gulf to Mexico, and some few make their way to the Caribbean – Cuba, Trinidad and elsewhere — while most stop in Florida and become reproductive. Some of the offspring of these – and of other individuals who over-winter along the Gulf coast in Louisiana or Texas – may, it's not known for sure, migrate back north in the spring.

Not all breeding monarchs leave their northern homes at the same time. Some appear to head south as early as July. Some of these pre-migration migrants, as Chip Taylor calls them, might stop in Texas or Oklahoma to reproduce. As many as 60% of females in the early wave of the big migration, arriving in Texas in early October, were found to be carrying spermatophores, while later arriving individuals were all in reproductive diapause.[76] A few late-emerging adults are found after most have already left the northern edge of their range, in mid-fall. They feel the coming chill and head off south on their own. But their chances of getting there are likely less than 10% of the success of the main group of migrants.

As mentioned earlier, the milkweed *Aesclepius syriaca* is the main nectaring source for southward migrants, *S. viridis* and *asperula* for those heading north from the southern U.S. Spring weather and the milkweed crop it brings, or doesn't bring, certainly affects the monarchs as they head north. A warm, wet spring will allow plenty of egg-laying, and provide good numbers to start off the summer populations farther north. The first north-winging migrants generally appear in northwest Louisiana and similar latitudes of Texas between the mid- March and early April. The migrants don't linger in the southern states; they breed and die, and the new generation develops rapidly and continues north. A breeding population doesn't stay behind — by late May they're no longer found in Texas or Louisiana. But what happens to their populations there is critical to the breeding summer population – a late heavy frost or other bad weather, or lack of milkweeds on which to oviposit at this time can put a serious dent in the later summer generations.

During the summer breeding season further north *A. syriaca* and *A. speciosa* seem to be the favored food.[77] When they arrive on their northerly breeding grounds of course varies from year to year. One observer in Shawano, Wisconsin reported first monarch sightings between May 21 and June 9 for the years 2003-2008 (Dan Everson, Dplex-1 list). Chip Taylor reports that the first monarchs usually arrive in the Lawrence, Kansas area at about the time of peak dandelion bloom, which happens to be when the first milkweeds are making their way up, and ready for eggs. In other words, no matter where you live, a flush of dandelions on lawns and fields is a good time to look for the spring's first monarchs.

As of the fall of 2007, nearly 400 tagged monarchs had been recovered after traveling more than 2000 miles to arrive at Mexico's over-wintering sites. These monarchs came mostly from Ontario and the northeastern states of the U.S. Most recoveries are at the El Rosario sanctuary, of tagged individuals who have died while over-wintering. The distance a monarch can fly in one day depends of course on the winds. If the weather's bad they need to hunker down and wait it out; headwinds, storms, rain can all cause that. With a light tailwind your average monarch might make more than 40 miles in a day. But one individual tagged in Ghent, New York on September 16, 2004 was recovered on September 24 in Jupiter, Florida. That's 1124 miles in

8 days or about 140 miles a day, a long trip for a human on a high-tech bicycle. Certainly not bad for half a gram insect.

In the summer of 2009 a large male monarch researchers called "Big Boy" was the first to fly any significant distance (more than 11 miles) carrying a tiny radio tag (Chip Taylor, Dplex-l list.) Thanks to this new micro-technology we're soon going to learn much more about the habits of monarchs as they flit through the air.

Monarchs gather in Mexico's oyamels because there the temperature is just right. Not quite freezing but cool enough to slow their metabolism and save their fat for the long five months waiting for spring. Because temperatures are not exactly the same year to year, it's also important they be able to move uphill if it's too warm (on average cooling themselves about 4 degrees Fahrenheit for every thousand feet of altitude) or downhill if it's too cool. This suggests the forests need to be left covering a range of altitudes. Among the oyamels they also find the mists which the locals call *el legarto* – the lizard which crawls away to hide when the sun comes out – the mists which provide the high humidity which keeps them from drying out, and nearby springs and pools to rehydrate themselves when they need to.

The canopy of the forest keeps them dry during rain, as does the cluster itself. They prefer south or southwesterly slopes. Slopes, so the coldest air will drain away from them, southerly to catch early spring sun to warm them should they need to fly for a drink. Like Goldilocks, they've found a spot which is "just right" – and as we will see in the next chapter, our Goldilocks are in danger of being evicted.

Since one to two hundred million or more monarchs from eastern North America come to gather in about 50 acres of forest, they achieve populations in the range of four or five million monarchs per acre at these sites. It is an awesome sight.

The oyamel forests of central Mexico might be just right, but they are not always just right. Hard winter storms, in the worst case freezing rain, can decimate their populations. It happened in January of 1981 when in just one of the over-wintering colonies more than two million monarchs froze to death. It happened again in the winter of 2001-2002.

Studies show that when dry, half of the adults can survive down to −7.7 C (18 F), but half of wet individuals die at − 4.2 C (25 F). Those three and a half degrees of centigrade happen to be right in the range of coldest temperatures at their over-wintering sites, so staying dry is really a matter of life or death. Wet butterflies fully exposed to the night air, without cover or cluster to protect them, are at special risk and may die at temperatures hovering right around freezing. Individuals who fall out of the cluster overnight can be fatally wetted by early-morning dew.[78]

Reproductive monarchs studied in Ohio were more sensitive to freezing than those who'd entered reproductive diapause and were preparing to migrate. Migrating individuals do not really show frost damage until ice crystals have begun to form in their cells. So even before heading south they've already undergone physiological changes which prepare them for a winter in the cold mountains.[79]

Over-wintering monarchs must be able to survive the cold mountain air while preserving their lipids. This requires sometimes warming themselves within the cluster, usually by fluttering their wings, and at other times avoiding warming, by moving out of the sun. Living as it is at the bottom end of temperature survivability, the over-wintering adult faces different challenges from the breeding adult who in high summer is more likely facing the threat of over-heating. Chip Taylor (Dplex-l list) has some evidence that smallest males & females, otherwise appearing healthy, die at somewhat higher rates while over-wintering. He suggests this may be because the higher metabolic rates of males and smaller individuals require breaking down more trehalose, a metabolic process which requires water. They may be dying of thirst.

The oyamel firs and the security of the cluster are refuges for the tired monarchs. But by gathering in such numbers the clusters become a rich resource to predators and parasites.

Two species of birds regularly feast on the banquet nature has brought them: black-backed orioles and black-headed grosbeaks. According to one estimate, over the winter they kill perhaps 15,000 monarchs a day, for a total of about 2 million butterflies over the course of their over-wintering, about 10% of the total population. The birds seem to prefer male butterflies, possibly because they have a higher fat content, and/or because they have lower cardenolide loads.[80]

Another study suggested that in 1979 these birds accounted for up to 60% of the mortality of over-wintering monarchs.[81] The two kinds of birds have different strategies for handling the toxins in the butterflies. Black-headed grosbeaks discard all but the abdomens, which are rich with fat, and seem to be able to tolerate the high level of toxins in the exoskeletons surrounding the abdomen. The orioles, with their sharper beak, slit open the abdomen and thorax and with their tongues scoop out the fat while discarding the exoskeletons. Biologists use these different dining strategies to distinguish which predator killed monarch bodies found on the ground. Death due to bird predation appears to go in an approximately eight-day cycle at the over-wintering sites. One suggestion is this is because it takes the birds about that long to break down the poisonous compounds they've ingested, before they can eat again. There is also evidence that over the winter the cardenolides slowly break down, leaving the monarchs less and less protected by them.[82]

A stereotypical image of Mexico is of a cactus-strewn desert with vultures circling skeletons. Vultures do not prey on dead or dying monarchs, but mice do. One study suggests at least five species of mice feed on the butterflies who've fallen onto the ground and froze, died of starvation or dehydration, or are just too cold to escape. *Peromyscus melanotis* is the commonest. Another study claims that species is the only mouse which eats monarchs, the others finding them distasteful.[83] Though there are more predators at Mexican over-wintering sites than at California sites, only about 10% of over-wintering monarchs in Mexico contain one emetic unit of cardenolides (that's enough to cause a blue jay to vomit), while about half of the California over-wintering population has that much cardenolide. One clear difference in over-wintering behavior between these sites is the tighter clustering of butterflies in Mexico than in California. While this may in part be because of somewhat colder temperatures, it may also act as a defense against predation.

Like predation, disease and parasitism also increase whenever populations gather. Over-wintering monarchs are beset with an especially troubling parasite in the protozoan *Ophryocystis elektroscirrha*. This pathogenic protozoan, like those which cause malaria or sleeping sickness in humans, forms contagious spores. These spores have been found on all parts of the monarch's bodies, though most commonly on the rear third of the abdomen. Individual monarchs with high spore count dried out sooner and had a higher mortality than those with lower spore counts.

A large-scale study of the prevalence of this protozoan on monarchs captured all over the U.S., not just over-wintering individuals, looked at nearly 15,000 monarchs between 1968 and 1997. A population of the butterflies in southern Florida, which did not migrate or disperse, had highest infection levels, with more than 70% of the individuals infected. Western populations which migrate but not great distances were infected to a level of about 30%, while eastern populations which travel to Mexico, averaged about

8% infection over those thirty years. But those which are infected, when they come into intimate contact with cluster-mates, spread the parasite, and as they over-winter infection rates increase. The biologists suggested that migration has two effects on parasite load – by dispersing the population it decreases parasite load, but by bringing the butterflies back together, it increases again.[84] The difficult migration weeds out the infected individuals and helps decrease parasite load. In one experiment some monarchs were artificially infected with the parasite. These individuals showed shorter flight distances, flew more slowly and lost weight more during flight than uninfected individuals.[85] It appears that adults collected at the over-wintering sites in California who were infected with the spores of this parasite dehydrated more quickly at low relative humidity than uninfected individuals.[86] And one study suggested that under conditions of high densities of larvae individuals tended to become more susceptible to this protozoan, with negative effects on their development. Even without the parasite, high densities of larvae affected development in harmful ways.[87]

A carefully-designed computer modeling of the evolution of parasite virulence – how rapidly and efficiently they reproduce in their host populations – used this monarch/protozoan pair. This study showed, as common sense suggests, that parasites that were too virulent killed their hosts too quickly and so were unable to spread. Parasites that were not virulent enough did not spread through a population. Protozoans which had intermediate levels of virulence were most successful.[88] These findings have the potential to improve vaccines for human diseases.

Migration not only takes the monarchs away from the cold snows of the north, into contact with others of their species from other locales, and allows a great springtime mixing of their genes, but it also works to bring down parasite loads. Evolution works best on adaptations which provide increased fitness in more than one way. As we are about to learn, this incredible biological phenomenon of monarch migration, shaped by evolutionary forces over thousands and thousands of years, but within the last few decades threatened by human activity, is in danger of extinction. Let's hope our children and grandchildren will be able to enjoy and marvel at the huge and lovely clusters of over-wintering monarchs, brought together by their finely-tuned navigational and fuel-conserving mechanisms.

Most cultures have a tradition of life after death, their own native eschatologies. Where that post-mortem life takes place, and what conditions one encounters there vary widely. To some it is a resting-place between reincarnations. To the Hebrews Sheol, later translated as Hades, was a dim gray tomb in which souls forever slept. Later versions of the after-life, incorporated into the Christian tradition, were influenced by the Greek conceptions of an underground fire-filled Hades and a much more pleasant Elysium. The Koranic heaven, home to martyrs who die defending the faith, is an eternal banquet of fleshly delights.

According to some, souls are not locked into their ghostly homes, but can wander back into our plane of being. There are different versions of why these spirits – *almas sanctas* — might want or feel the need to return. To the classical Romans one type of spirits were the *lares*. Family lares were spirits of the ancestors, deserving veneration. Another word referred to the souls of those who'd died a violent or early death and came back to this plane because they hadn't had full opportunity to move on to the next. These they called *larvae*.

Are the monarchs *almas sanctas*? Are they the returned souls of the dead, as the indigenous people of Mexico's central highlands were taught from their aboriginal religions, and many continue to believe under the umbrella of Catholicism? Why have these spirits come back? Is their long over-wintering a kind of purgatory? Do they fly down out of the oyamels — named by Alexander von Humboldt the *Abies religiosa* because their end twigs are shaped like a cross — do some fly down out of their cluster to drink from the waters of Lethe, the waters of forgetfulness, to forget and in that way finally be allowed to escape their past? Or are the waters those of Mnemosyne, which according to the Orphic tradition remind you of who you truly are, and in so doing bring you to a kind of enlightenment which releases you from the round of reincarnation?

These belief systems, call them fictions if you like, are like a well-crafted story, first told and retold, then written and rewritten many times. They are very lovely. It is certainly fair to question their literal truth, but it's also important to remember how they reveal deep truths about ourselves, and important truths about our relationships with other of life's creatures, relationships more ancient and meaning-full than those encountered in the chat-rooms of the internet.

Our emotional connections to butterflies and what they represent to us go very deep. So they show up in children's art, even in concentration camps. Harriet McDonell, from the tiny town of Glen Flora, Wisconsin, shared with me how butterflies are so important to her – and she, to them. Though disabled, she has a lovely flower-garden in which she spends summer days. There she communes with the butterflies. If there are none she calls them in;

and they usually show up. Their lithe beauty comforts her. They provide her a quiet company. Harriet shared a more personal story with me. She happened to be in the hospital at the same time as her elderly mother. Harriet was on the fifth floor, her mother in intensive care on the second. From her window, Harriet could see down into the room where her mother lay attached to oxygen and the various monitors one encounters in critical care. After a day or two, Harriet noticed a monarch butterfly flitting about her window. When her son came for a visit, she told him that the butterfly was her father, who'd died some years before, come to get her mother. A few days later there were two butterflies outside the window. Looking down into her mother's room, she saw that her mother's eyes were open and unblinking. She told the nurse's assistant that her mother had just died. She had.

There is much that science can tell us about the world. There are truths – the truths of poetry, art, religion – which may or may not possess an objective correlative, and whether or not they do may not be that important. Those who insist these stories meet the same criteria of truth as astronomy or chemistry miss the point as much as those who insist they are actually of the same order of truth as scientific truths. These other truths are true according to different criteria, because they describe a different world, the world of the human imagination.

Wallace Stevens' poems are maps of sorts of the worlds created by human imagination. He says of the woman who sings by the sea in *The Idea of Order at Key West*

"She was the single artificer of the world
In which she sang. And when she sang, the sea
Whatever self it had, became the self
That was her song, for she was the maker..."

The sea, the moon, the squawking blue-jay and our friend the monarch are winged creatures who offer us a ride, to destinations unexplored, for the essence of the creativity they offer is to take us into lands as yet unknown, as yet uncreated. As much as science has taught us about the sea and the moon and the butterfly, to deny them this potential is to deny them the chance to teach us something about ourselves. And what a pity that would be.

But the story of what science has taught us about our fellow-citizens the monarchs is to me as lovely as those other more imaginative truths, and in a way even more wonderful, not only because its truths have been verified, but also because we can fit each tiny truth, gained through moments of inspiration and hours of work, into the larger picture of how species relate to one another and we to them. It is an eschatology of its own, pointed not to some far-off place but right back down to the terra firma at our feet.

Chapter 5
Conservation

"But evil things, in robes of sorrow,
Assailed the monarch's high estate.."
from Edgar Alan Poe, *The Haunted Palace*

I bear good news, and bad.

First, the good.
Step off the planet for a moment, pretend you're a god or demigod thumbing through the long sweep of human history. You'll have to admit we humans are a remarkable addition to the earth's biodiversity. We write supremely delicate and lovely poetry, in even the most appalling circumstances. We craft the finest pottery and sculpture, and raise buildings as beautiful as they are functional. We weave fabrics into astonishing patterns, fabrics too of myth and spirituality rich with meaning. We design and build complex machinery, tools with which we reshape and now reach beyond the earth. We dive – sometimes belly-flop – into the hidden mysteries of nature. Some of us have begun to feel the threads of those mysteries as connections to other species. We are, in short, a creative, curious and competitive creature.

Now the bad.
From another viewpoint the sweep of human history is not all that reassuring. We've spent far too much time and energy mistreating one another, in very big and sometimes little ways – other humans as well as the earth's other species. We invite ourselves into their homes and declare them and everything we find in them to be ours. We roam the earth with big machines, swooping life up out of the ocean and knocking it from the air with lead, our cars and trucks or cell phone towers. As we move across the globe on pleasure and business we bring some as uninvited guests who like us are only too ready to declare the territory their own. We drop our wastes onto the earth and into our waters and throw them willy-nilly into the air, changing the landscape, perhaps even the climate.

In all these ways we have touched monarchs too. Our clearing of the primal forests of the Eastern and Midwestern U.S. two hundred years ago affected many species in many different ways. At first it likely helped monarchs, by converting large tracts of forests to fields and meadows. Now many of those meadows and pastures have become sprawling fields worked by gigantic tractors, some driven not by human hands but automatons linked by GPS to satellites, noisily spraying those fields with the products of our creative chemistry. We challenge these gentle half-gram fluttering creatures to thread their way through a gauntlet of one highway after another thick with screaming steel. As though that were not enough, we have torn down and usurped their over-wintering sites, once thousands of acres, leaving them only small remnants of their forest refuges.

Paraphrasing Aldo Leopold, if you're going to care about nature be prepared to have your heart hurt. If you care about the monarch, steel yourself against the rest of this chapter. It is a bitter pill, sweetened only a little by the hope that compassion and understanding might win out over the larger forces of population growth and consumerism, bent on carving our name in big, ugly letters across the landscape.

We are timorous bullies, the worst kind, our hard-swinging aggressiveness driven by a deep-seated insecurity. It is no surprise, this insecurity, when you consider the larger forces of nature around us, much of it even now inexplicable, unpredictable, dangerous. E.O. Wilson has described us as a biophilic species, loving our fellow creatures. Yes, we are. We love our dogs, our cats, our pet parrots, the birds that come to our feeders. But not all creatures. Not those slithering through the grass which strike out, bite and kill us. Not the tiny spiders who can do the same. Not the crouching tiger or even the 900- pound gorilla who, but a gentle herbivore, has the temerity to protect his family. We love long vistas from the security of our houses or cars; we love our parks and rivers, so long as we feel safe in them.

Cave-bears were a very large species of bear, larger than grizzlies or polar bears, which went extinct at the end of the last ice-age. According to some historians, it was the accidental discovery of their skulls in the caves of Europe in the 16th century which jump-started our fascination with dragons. Until then only rumored, these fire-breathing creatures who lived in caves and threatened civilized life were brought to life by cave-bear bones, and also the bones of the extinct woolly rhinoceros, paraded from town to town as dragon bones. Though the dragon myth had idled along in the background of Western culture for millenia, like a shot of adrenalin the discovery of these skulls inspired poets and painters of the time, when dragon-art suddenly became popular.

Evidence suggests that dragons are actually the invention of more oriental cultures. In these Eastern cultures, such as the Chinese, dragons are powerful but mostly friendly creatures whose help we humans depend on in our struggles to survive. But in the West dragons instead represented the dark side of nature, lurking in the thick forests beyond the pale. The male hero – Saint George or John Waye — stands up against that evil, and protects the sanctity of our softer, kinder, but more vulnerable feminine nature.

Among other things we are a symbol-making species. We fill our lives with and find meaning in flags, trade-marks, and religious symbols. For those of us raised in Western culture the dragon is a symbol of wild nature. It is an understandable response. Children wandered away from home and were taken by lions, or tigers. Others lost their mothers to snake-bites, floods, and storms. Some still do. Wolves took lonely travelers lost in the deep forests of northern Europe. It was a jungle out there, and in some ways it still is. A recent newspaper account described a boy dragged out of his tent in Utah and killed by a black bear, another of a young woman attacked and killed by coyotes in Ontario. These attacks are now rare, but for most of our history our species has been prey as well as predator. The fear of darkness, thick woods, and strange noises in the night is not irrational. I write this on Halloween, our national celebration of the dark side, when we try to trick our fears by dressing up as them.

I've yet to find no satisfying explanation for the black & orange of Halloween. The black of course for death and darkness. But why the orange? Some say pumpkins. Some say it's the color of fall, and harvests. Is there, I ask myself, any connection to *el dia de los muertos* in the hills of Mexico, our Halloween, when the orange and black souls of the departed reappear?

Some of our fears are rational – the fear of death and of the monsters in our own psyche which now and again rear their heads and strike. But many of these fears are, after all, only monsters of our own making, dim insubstantial shadows from days long gone which like unchecked endoparasites or ghoulish vampires derive their life-force by stealing from ours. It is time, I think, for a larger perspective. Time to ask ourselves, in another but very apt analogy: are we really willing to sacrifice all those innocent bystanders in the name of our own, homeland, security?

Because of us, many species have gone extinct, and continue to go extinct. Like the loss of innocent lives which happens in warfare, it is not a pleasant thing to consider — but does turning our backs on the reality of it make it any more pleasant?

Some species are sensitive to ecological change or disturbance, others more resilient. Endangered species usually share a suite of biological, genetic and ecological characteristics that put their populations in peril; traits which, fortunately, monarchs do not possess. Instead they are a wide-spread, opportunistic and adaptable species we are not likely to drive to extinction. But in 1983 the IUCN – the International Union for Conservation of Nature and Natural Resources – designated their remarkable migration as threatened. Though various human activities threaten this migration, the most urgent issue is the continuing destruction of their over-wintering sites. If those sites disappear, as well they might, yet another of the Wonders of the Natural World will be gone.

Amalgam a deep-seated insecurity in the truly wild with human brains, brawn and technology and you have the story of Western culture's domination of nature. In Mexico one word for wilderness – wild lands — is *maleza*, a sibling word to malaria, malady, malice and malignant, from the Latin *malus*, meaning bad. The language itself tells you the wild is not just "wild," it is bad. So it is no surprise that the wilderness is destroyed, regardless of the fate of our monarchs. And it is no surprise either how hard it's been to convince those who scrape out a living near the forests, and the politicians who represent them but even more the better-heeled logging interests to preserve that wilderness. Perhaps the real surprise is that their attitudes show some evidence of changing.

Mexico is dotted with variously-aged volcanoes which like fairy-tale giants wake now and then from their long sleep to stretch and shake up the countryside before slipping into another long nap. The particular volcanic range most monarchs fly to is called the Transverse Neovolcanic system, located in the states of Michoacan and northwest Mexico, about four hours by bus west and northwest of Mexico City. The slopes of these mountains are covered with oak, pine and fir forests. The preferred habitat of the monarchs lies between about 2500 and 3500 meters (about 10,000 feet) altitude in the oyamel firs (*Abies religiosa.*) The oyamel fir forest is a unique ecosystem, a remnant boreal forest found only in thirteen "eco-islands" in mountainous Mexico. Nine of these sites are in the Transverse Neovolcanic system itself.[1] These fir forests are surrounded and often dissected by juniper shrublands and grasslands, and sometimes crop and pasture-lands.

Because it is primarily in the oyamel firs that the butterflies cluster, preserving the mature firs is essential to conserving the over-wintering butterflies. It's not just a simple matter of not cutting all the trees they cluster on. Even thinning the forest increases mortality of the monarchs from freezing, due to increased heat radiation to the night sky and the increased likelihood of their wetting. Thinning also increases winds in the forest which shake some of the butterflies out of their clusters, where they become prey to mice.[2] Bird predation in areas of the reserves which have been logged has been shown to be higher than in closed canopied, undisturbed areas.[3] And as we ecologists are fond of saying, a thousand trees do not make a forest. A forest is an entire, complex ecosystem, in many ways more complex than a car or computer. The actual number of acres the butterflies cluster in is incredibly small; less than fifty, total. It is not only the oyamel forests themselves which need preserving, but also the mix of habitats naturally found in and around them. Because they've hopefully arrived full of lipids, the monarchs generally do not eat while over-wintering. In fact there is little food for them there anyway. But they do dehydrate. Thinned forests have higher wind velocities, which likely dehydrate the monarchs faster. While flying to get a drink they use up their lipids much faster than while clustered, and this can lead to starvation, or reduced reproduction. There is evidence[4] that in drought years the monarchs leave the reserves earlier than usual. That can cause them to arrive in the U.S. before the milkweed is up, which affects their ability to reproduce.

Standing in an over-wintering site on a sunny winter day, with the monarchs streaming around you down the hillside for a drink is an unforgettable thing. To them it's a serious business, this finding water. They get some water from dew which forms on the vegetation around them; perhaps over the course of the winter, the source of most of it. But they also find water in seepages from groundwater reaching the surface, and along watercourses. These water sources are an important resource to the over-wintering monarchs. Without them and the cover around them, the reserves cannot support nearly as many butterflies. That means the total area we need to preserve is considerably larger than just the forests themselves.

Once discovered, efforts began to protect the over-wintering sites with their millions of clustered monarchs. Mexico's then-minister of agriculture and forestry, Cuauhtemoc Cardenas (who later ran three times, unsuccessfully, for president) worked to protect the monarchs. An organization called *Pro Monarca* was formed. To publicize the effort, the government issued a postage stamp and a coin with monarchs on them. The popular poet Homero Aridjis, who'd grown up near one of the proposed butterfly reserves, took up the cause. "For me," he wrote, "they are visual music, the apotheosis of light, motion, and beauty, a solar-light symphony. In the morning the butterflies are in the trees, with their wings closed. When the sun comes out it touches their wings and they begin to open. Noon is the hour of activity. I learned to love nature through the butterflies." Aridjis has since dedicated himself to the conservation of three of Mexico's most charismatic migratory species: monarchs, sea turtles, and the gray whale.

Rodolfo Ogarrio, a long-time proponent of the monarchs and lawyer who runs the Mexican Foundation for Environmental Education, describes first going to see the over-wintering monarchs: "It was something that changed our lives. It was a perfect blue day. All the butterflies were flying because of the bright sunshine. We had no information or understanding of anything...The experience was sensory, almost religious. I remember this scene of us lying on our backs and watching the butterflies, a silent river of wings, fly down to a small creek to drink, thousands of living pieces of stained glass against the bright-blue sky moving in all directions. We stayed hours in silence, not speaking, absorbing something we realized from the very beginning was sacred...I left with the thought, this has to be protected."

The effort to save the monarchs' remaining over-wintering sites has been difficult from the start. A single large fir tree can bring $300 or more, enough to feed a family for a year. Locals want access to the reserves to gather firewood, herbs and roots. Loggers want jobs; logging companies want profits. The federal government in Mexico has bigger "fires" to deal with than the preservation of monarch habitat – at least from their point of view. Drug cartels bring violence and graft into a government never known for its transparency. Within the past few years human heads have been left as messages to governmental authorities not to push the cartels too hard, some right in Zitacuaro, one gateway to the monarch reserves. Some tours of the Biosphere Reserves originate in the city of Morelia. On September16 of 2008, Mexico's Independence Day, seven bystanders were killed when grenades were tossed into Morelia's busy plaza.

The loggers responded to the declaration that the forests would be preserved by rushing in and clear-cutting many acres – as a form of protest, to chase away the butterflies threatening their jobs, as well as getting their hands on the timber while they could. An additional threat has arisen in the form of a kind of terrorism in and around the Reserves – what appear to be purposeful fires as a kind of protest to the conservation of the Reserves. As of this writing these fires have not threatened the actual colonies of monarchs.[5]

Mexico is a very poor country. Something like half the people live on less than a dollar a day. Unemployment is endemic. The Mexican government, like many third-world governments, is torn by the countervailing forces of development and preservation. Money drives politics, and there's more – or at least quicker — money in development than in preservation. So the politics can get complicated and ugly, and often does. Deforestation of the Reserves has effects on locals who live in the area. As the trees are cut down, streams which provided the monarchs water dry up. Or, as happened in February of 2010, deforestation followed by heavy rains can create ecological and human disaster. The city of Angangueo, gateway to the El Rosario Reserve, was nearly destroyed by catastrophic floods and mudslides, leaving 80% of the residents homeless and dozens dead. As of this writing, the effect of this disaster on the monarchs themselves was unknown. The sound of chain-saws spills sometimes down off the mountains, supposedly protected. Armed posses gather to do what the government seems uninterested in, or incapable of, doing – enforcing the laws.

Sometimes the government does step forward. In December of 2007 nineteen clandestine sawmills in Michoacan were raided by police. Enough illegal logs taken from the buffer zones around the reserves were confiscated to fill 400 logging trucks. Some believe that rather than prevent the logging, the government allows it to happen, confiscates the logs, and in selling them enriches their own pockets.

In March 2004 a fleet of logging trucks moved illegally into the Sierra Chincua reserve and in the fall of 2004 on the slopes of El Pelon. A hundred armed men confronted ten policemen when they attempted to stop logging trucks in the Angangueo region in September of 2005. In a March, 2008 New York Times story, Lincoln Brower and others report discovering through careful study of NASA photos of an *ejido* (land-holding) called Crescencio Morales, that significant logging occurred (more than 700 acres total in the previous two years) in the Core Zone of the Reserves.[6] A news article in December of 2008 reported that more than a million dollars' worth of logs had just been seized from illegal logging operations. That the forests in the reserves will actually be saved is never guaranteed. Preserving them is part of a constant, and continuing, battle. In the winter of 2008-2009 local residents were sometimes engaged in fights with loggers in the Reserves. Smaller-scale logging is harder to stop, and continues to damage the Reserves (Lincoln Brower, Dplex-l.) The Mexican government, seeing the potential for ecotourism dollars, in a choreographed media blitz announced in the fall of 2007 it would invest nearly $5 million in improving protection of the monarchs. This may be little more than window-dressing.

Organizations like Mexico's *Pro Monarca*, the *Lepidopterists' Society* (and its Mexican equivalent, the *Mexican Society of Lepidopterists*), and the *Xerxes Society*, an international organization which focuses on the conservation of invertebrates, fight the fight for us. The Mexican NGO *Pronatura* has identified six species of special concern in Mexico, including the monarch. Many individuals from many organizations and countries have contributed time, effort and money to the cause. They are too many to name here; they are the unnamed heroes of this cause. They need your help.

Over-wintering colonies of monarchs have been seen outside the Transvolcanic sanctuaries.[7] There have been unverified reports of colonies near Oaxaca and San Cristobal de las Casas and in fir forests in Guatemala. Large flocks were seen migrating near the border of Guatemala, headed northwest. Flocks have also been seen flying south in November through the Sian k'ian Biosphere Reserve south of Cancun. These and the Guatemalan butterflies might be eastern individuals who migrated south through Florida and crossed the Caribbean into Mexico. These outlying groups are thought to be temporary colonies which most likely do not return to the same sites year after year, and deserve more study.

There are reports of over-wintering sites on Mexico's volcano Popocatepetl. Monarchs are great opportunists and if the habitat is near right, they'll likely use it. Popo plays another part in monarch destiny. When it erupts it sends up huge clouds of ash and dust, which can cool the local weather. That happened in 2002 and likely contributed to the big die-off of monarchs that winter – perhaps as many as 200 million, or more.[8]

Most years there are some twenty colonies of over-wintering monarchs in the Transverse Neovolcanic system itself. In the 2009-2010 winter, only seven colonies were found. The summer 2009 breeding population in the U.S. was unusually low, and one estimate was that over-wintering populations would cover between 2 and 3 hectares in the 2009-2010 season (Chip Taylor, Dplex-l list.) Actual population measurements at the reserves that winter suggested the butterflies only covered 1.92 hectares, an all-time-low since measurements began going back to 1975. In the short-term, Dr. Taylor suggests that hot, droughty conditions providing little food for the first spring generation of larvae in the southern U.S., combined with dry and cold conditions in the upper Midwest in the summer of 2009 likely limited population growth. But the mean colony coverage in the reserves for the 1990s was about 9 hectares, for the first decade of the 2000s only about 5 hectares. This long-term trend is troubling.

Monarch populations experience large swings. That annual variation is due to a number of factors – how well the colonies over-wintered, whether there are young milkweeds for them to deposit on as they head north, how many generations are raised by the more northerly populations, which itself is determined by the arrival date and local weather, etc. Add to this variation in parasite or predator population and human-caused factors such as loss of habitat, pesticides, etc. and it's clear the equation determining each year's population size is complex. Long-term studies of population trends are critical to understanding the interplay of these factors.[9]

After a decade of hard work filled with conflict and controversy, some of these critical few acres were finally designated in 1986 as Biosphere Reserves. But this was only a small victory in a small but critical battle, while the war goes on. One study compared aerial photographs of the reserves and surrounding countryside taken in 1971 with those taken in 1994 and 2000, and fine-tuned the results with an on-site plant inventory. Of the 2000 acres of new farmland created by deforestation between 1970 and 2000 in that region, 200 acres were within the designated monarch Reserves. The good news is that the deforestation rate in this area is only 0.1% per year, while in the surrounding regions it is seven times higher. The not-so-good news is that an additional nearly 4000 acres of the Reserves, though not deforested, shows evidence of disturbance from timber harvesting, firewood collecting, plant harvesting, and other human activities. Recovery of previously deforested areas through tree-planting was seen on about 2000 acres, and reforestation of farmlands on about 2500 acres. But all these recovering forests had small trees which would remain essentially useless to the butterflies for decades. Some reforestation, such as at the La Cruz Habitat Protection Project, was meant to provide locals with plantation wood as an alternative to cutting down the fir forests.[10]

Another study of aerial photographs showed that between 1971 and 1999 about 44% of the monarch Reserve forest showed signs of being degraded. The largest patch of high-quality forest was reduced from 27,000 hectares to less than 6000 hectares. While the previous study found a deforestation rate within the Reserves of only 0.1%, this study concluded that the rate of degradation of the forest in the 15 years preceding 1999, which includes all human impacts on it, was 2.4% per year. These authors concluded that the 1986 decree was only marginally successful at protecting the forests.[11] In the year 2000 new boundaries were established to the Reserves which (one hopes) will provide additional protection for the butterflies.[12] But protecting the monarchs remains a difficult and sometimes dangerous job.

Preserving these Reserves is made more difficult by Mexico's poverty, the government's efforts at increasing industry and exports, even the language itself. Another issue is the complex history of land ownership in Mexico. The government has purchased some portions of the Reserve forests, which provides them maximal though still fragile protection. Within the Reserves lands not owned by the government are owned by a combination of 59 *ejidos*, 13 indigenous communities and 21 private owners. About half of all land in Mexico remains in *ejido*. This type of land ownership is unlike anything in the U.S. or Canada. It dates back to the Mexican Revolution of 1917. Previous to then, much of Mexico's land had been arbitrarily granted by land-grants to wealthy, well-placed supporters of the emperor. Rather than owning the land they worked, local peasants worked it in a semi-feudal arrangement. They were allowed to keep a portion of their harvest for themselves but provided the biggest part to the landlord. Perhaps the most important issue driving the 1917 revolution was agrarian reform. After its success large tracts of land were taken out of the hands of the landlords and granted to the local communities rather than to individuals. Though they didn't actually own, and therefore couldn't sell their land, farmers could now work the land and keep their entire harvests. Unfortunately for preservationists, they were required to work the land – which meant cutting down the forests and planting crops or pasturing cattle. According to this system, similar to mining rights and the old homesteading system in the U.S., the peasants would lose control of un-worked lands. In the 1930s Mexico's Franklin Roosevelt, Lazaro Cardenas, granted another 50 million acres to *ejidos*. At that time many of these *ejido* properties were organized into collective farms, adding another complicated layer of ownership and control over the lands.

In 1992 a new law was passed which allowed *ejiditarios* – those who worked the collective farms — to obtain ownership of some part of the *ejido*, by going through a lengthy legal procedure. This complicated system of land ownership and transfer of properties has significantly slowed down the protection of the Reserves. While attorneys for *ejiditarios* on the one hand and local government and preservationists on the other work out details of ownership of the forests, legal and sometimes illegal logging and clearing of these trees continue.

Now, with growing national and international interest in the Reserves, and the experience gained over several decades, the efforts of the preservationists have slowly become more effective. Preservationists have adopted a multi-tiered strategy. Their work now includes lobbying land-use issues, administrative protection of the reserves, ecological and entomological research, forestry, socioeconomic development, education, construction of facilities, and tourism. Through educational efforts directed at governmental authorities, local residents and eco-tourists, they've changed perceptions of the butterfly and its habitats. Their financial support for research on monarch biology and especially the effects of forest practices on the over-wintering populations is based on the assumption that the more we know about the over-wintering monarchs, the better we will be able to protect them.[13]

The most important strategy for protection of the monarchs is the protection — and, eventually enhancement — of the oyamel fir forests. The present forests are relicts of much larger tracts which spread over central Mexico as glaciers moved south over the central U.S., cooling Mexico. About 10,000 years ago the glaciers retreated for the last time and as the climate warmed temperate or even subtropical ecosystems moved in. Only remnants of the boreal fir forests remained, at higher and therefore cooler altitudes. At most there are about 100,000 acres of oyamel firs in all of Mexico, at about 8500 - 12,000 feet altitude. Postglacial warming, at the same time it pushed back the margins of the oyamels, also allowed first milkweed and then the monarchs to expand northward. Over time the evolutionary play of monarch migration came to be written on this ecological theater.

So the biggest threat to the oyamel forests is from logging. The World Wildlife Fund determined that within the Reserves, between 2001 and 2003 illegal logging occurred in some twenty-eight communities, of which twenty-three were in the heart of the Reserves where all logging is illegal. As observed with many other species, individuals which reside on the edge of a habitat suffer a higher risk of predation than those farther in from the edge. So as humans chop larger tracts of forests into smaller pieces, this exposes more monarchs to predators.

Some argue that logging the oyamel firs will create open spaces for flowers. While over-wintering a few monarchs do leave the cluster to feed on scattered nearby flowers. These individuals are so low on lipids they are essentially starving. Since it is unlikely they would be able to renew their lipids to the levels necessary for reproduction or the northward migration, most of the monarchs visiting those flowers won't contribute to the next generations. Maintaining the forest canopy for the healthier monarchs is more important than creating openings for flowers.[14]

The federal response to the request for help with enforcing the no-logging rules has been disappointing. Local preservationists believe that effective protection will require a permanent presence of the army, destruction of the access roads, close monitoring of the forests and sawmills, and the beginning of a long-term, effective dialogue between the communities, law enforcement and loggers. Whether this happens or not will likely determine the fate not of the monarch as a species but of the remarkable phenomenon of their mass migration.

A tri-national effort (Canada, Mexico and the U.S.) to address these thorny issues, signed into formal agreement in June, 2008, is the North American Monarch Butterfly Conservation Plan or NAMCP. The July, 2008 announcement that the monarch Reserves were designated by UNESCO as World Heritage sites is another hopeful sign. A recent review article summarizes the pressures on the monarch Reserves.[15]

From what is considered roughly issues of most to least concern, other threats to the Reserves besides logging include fire, disease and pests (such as bark beetles, the geometrid moth and the parasitic dwarf mistletoe), live-stock grazing, and tourism. Both logging and fires (accidental or otherwise) usually result in conversion of the forests to agricultural land. Even selective thinning of the mature firs changes the microclimate of the forest, rendering it unusable for the over-wintering monarchs. While eco-tourism does help protect the forests it brings other impacts. Access roads and walkways can cause erosion, and locals who sell souvenirs or food to tourists sometimes cut trees down for firewood. Most of the people who live in or around the Reserves are at subsistence level. Preservation of the butterflies isn't as high a priority for them as it is for the wealthier eco-tourists.

This is how Homero Aridjis describes the conflict over the Reserves: "Year after year we see the same depredation. Year after year, the oyamel forests get smaller, placing the migratory phenomenon at risk. The problem lies with the greed of the loggers, some *ejido* members, and politicians who have formed ominous and corrupt alliances to sabotage the Reserve. As long as the greed and complicity exist the monarch sanctuaries are endangered. Only a permanent and armed vigilance in the Reserve, enforcement of the law, and a citizenry determined to defend the presence of the monarchs in the oyamel forests of Mexico can combat this tendency."[16]

In short, what can be done? Better forest management practices in forests outside the Reserves could replace all the timber gained by logging the Reserves, and the lumber and firewood needs of those living around them. To accomplish this will require long-term effective negotiations between the *ejidos* in or near the Reserves and those farther away. New cottage industries need to be established to replace income lost from logging. Christmas-tree plantations, already established in the area, are one option. Improved crop management strategies which increase the per-acre yield on agricultural lands outside the Reserves could replace crops grown in the Reserve. Population control must be part of the solution; if the nearby villages continue to grow, it likely won't be possible to provide them with resources without destroying the monarch sites. Fortunately, for the last thirty years or so Mexico's fertility rate has been declining. Only time will tell whether that decline, not equal throughout the country, will occur fast enough to save the monarchs.

We are not the only species to threaten the oyamels. Mistletoe is a parasite on other plants. As it grows it can kill its host. Control of mistletoe within the Reserves will require creative silviculture, supported by legitimate research. Pesticides used to control geometrid moths, which attack the firs, have the potential to damage the monarchs. Dipel, the commercial name for *Bacillus thuringiensis* (*Bt*) has been used on some of the forests. It's effectiveness has been spotty, and there is always the potential of unwanted side effects to the monarchs. Control of the bark beetle, another pest of the trees, will require proactive strategies like using pheromone traps or quickly culling any trees which do become infested.[17] In October of 2009 it was reported that some 10,000 infected fir trees were being cut down to slow the spread of this pest. Bark beetles, which have destroyed large swaths of pine forests in the American West, are a real threat to the monarch's refuges. It appears the oyamels are stressed, perhaps due to global warming and/or regional droughts, and it is the stressed trees which are most susceptible to the beetles. But so far only about 200 acres of the 30,000 or so acres of the Reserve have been infected, and the 10,000 infected trees represent a small portion of the 5 million or so total trees.

Of course it is not just the Mexican Reserves that need conserving. As pointed out below, even on their breeding grounds, and certainly on their pathways to over-wintering sites, monarchs run dangerous gauntlets. While some migrants take a more easterly pathway the main central route to Mexico seems to be the one used by the vast majority of migrating monarchs, and so deserves most conservation efforts.[18]

The information-base for effective conservation is building, as are the techniques which make use of this information-base. To guarantee the long-term survival of the oyamel firs, and of the phenomenon of over-wintering monarchs, will require hard work, money, and mostly political will. Only time will tell how serious or successful our efforts will be. You can help by sending any contribution you feel able to make, to the organizations listed above and at the end of this book. You can also help in another, very pleasant way, which is to travel to one of the monarch Reserves. Let the hotel you're staying at and the restaurant you eat at know you bring your money to the region because of the monarchs. It is an unfortunate truth that dollars speak louder than poets.

Human conversion of the earth's landscapes to our own uses, especially agriculture, continues. The human population, especially in developing countries, continues to grow. One impact of global warming on the monarchs is the possible loss of the oyamel fir forests themselves.[19] But weaning ourselves off fossil fuel and onto biofuels is itself a clear threat to natural communities in the coming decades. Already about 37% of the earth's available land mass is used for agriculture, predicted to increase to over 50% by 2050. The biodiversity impacts are staggering. The British have documented very significant declines in bird biodiversity on farmlands over the past 35 years. Even seemingly "green" energies have their impacts. For example, it's not known what the effect of large wind turbines, sprouting up all over the landscape, will be on butterfly populations. Specific design details, such as how tall they are, how fast the turbines rotate, and so on, can make a big difference.

The authors of a recent paper developed a simple risk assessment strategy for estimating the impact of a particular agricultural practice on local biodiversity. Perhaps a similar methodology could be applied to monarch conservation.[20]

The extinction of the passenger pigeon early in the twentieth century was not only the final extinction of that bird but also of a natural phenomenon, whose vast flocks of up to a billion birds were a spectacular part of the uniquely American natural landscape. The phenomenal migration of the monarch is likewise threatened, due to the small and diminishing size of the over-wintering sites in Mexico as well as California. At least twenty-one of California's over-wintering sites have already been destroyed, with an additional seven or so threatened. At Montana de Oro State Park, near San Luis Obispo, attempts at destroying non-native eucalyptus trees have already damaged a large over-wintering site there.[21]

Though there has been some success at preserving monarch over-wintering sites in California — Ellwood Mesa near Santa Barbara is an important example — Americans do not stand on high ethical ground when preaching conservation to others. Only continued efforts by everyone who cares will prevent the collapse of the oyamel fir forests and the likely collapse of the migrating populations.

While the greatest threat to the phenomenon of monarch migration is the loss of their hibernation sites, lets look at some of the other reasons to be concerned about monarch populations.

Working feverishly to create ever-new chemicals to combat the insect pests which would take our crops from us has had little effect on the proportion of potential harvests consumed by insects, which has remained for the past forty years at something like one-third of each year's yield. Application of pesticides has side effects on non-target species such as monarchs. These side effects go beyond just killing the non-target species. Sub-lethal effects include physiological, developmental, immunological, reproductive, behavioral effects and decreased longevity.[22, 23] All these things affect the species' ability to maintain a stable population. About 2 million pounds of the pesticide permethrin is sprayed on the U.S. landscape, half or so applied to wetlands to kill mosquito larvae. Recent studies suggest it can be very toxic to monarch larvae.[24] Because some species of milkweed grow in and around wetlands, spraying of permethrin is undoubtedly affecting monarchs.

The sale of agricultural herbicides in the U.S. is a multibillion-dollar industry. In California alone some 200 million pounds of pesticide are introduced to the environment annually, a $30 billion or more annual business. For the entire country, the numbers are much greater. Herbicides are applied to kill non-crop plants, such as milkweeds or other nectar sources for adult butterflies. They do so at increasing efficiency, and surely by doing so – and through other unintended side effects — impact monarch populations.

As we learn of the human-health effects of common pesticides we search for new, less harmful varieties. One of the most promising was the use of actual bacteria to kill insect pests. A species of bacteria called *Bacillus thuringensis* produces a protein which was found to be toxic to insects but essentially nontoxic to mammals. This is good. Kill insects – in fact only certain targeted insects – with a chemical not toxic to humans. But enter the advent of genetic engineering, and the plot thickens. Scientists have cut the gene for making the insecticidal protein *Bt* out of the bacteria and inserted it into various crops, including corn. That means when the crops grow they make that protein. In other words, they make their own pesticide. Nifty, huh?

Bt was first introduced into tobacco and tomatoes. In corn it is used primarily against the European maize borer, a lepidopteran like the monarch. One of the advantages of *Bt*, played up by Monsanto, main producer of *Bt*-engineered corn, is that only those insects that actually attack the corn would be damaged by the *Bt*. Since our monarchs don't eat corn, it would seem they should be safe. Then Losey *et al*, working at Cornell University[25] set off a fire-bomb with a short paper showing high levels of mortality in monarch larvae who ate not the corn leaf itself but milkweed leaves dusted with the pollen from corn engineered to carry the *Bt* protein.

Since milkweed grows as a weed in cornfields, there is the potential for corn pollen to contaminate this favorite food of monarch larvae. Studies suggest milkweed densities within agricultural fields is actually higher than in surrounding non-agricultural areas. As they're headed north through the midsection of the U.S. spring migrants can't help but pass through and over the millions and millions of acres of corn, acreage which is likely to increase as corn-based ethanol catches on. It is reassuring that in early summer, when the migrants are passing through, the corn is too young to produce pollen. But the summer reproductive generations also spend time in cornfields, so the concern seems justified. The release of the Cornell researchers' results led to an outcry in the press about the use of *Bt*-corn.

The swift reaction to the Losey *et al* paper was based on the Precautionary Principle: "Better safe than sorry." When we apply this principle in our everyday lives we sometimes err on the side of caution, and quickly forgive ourselves. But when environmentalists raise concerns about an issue and those concerns later turn out to be unfounded, the anti-environmentalist media is unforgiving. Alar, a plant growth hormone sprayed on apples to help them ripen, was banned decades ago because of potential toxicity in apples. Saccharin, a sugar substitute, was found to cause certain types of cancer in rodents, and removed from the market, though later studies showed human effects would only show up after massive dosages of the substance. These "false positives" are dragged out time and again in an effort to discount other environmental issues, such as global warming. The interplay of science and policy is a complex dance. We need to use the best science available when making policy decisions about environmental problems. But we should also remember that "Better safe than sorry" is always going to bring with it a certain number of inevitable false positives. Does that mean we should discard the Precautionary Principle?

The Losey *et al* paper on the potential effects of *Bt*-corn on monarch larvae is a good example of how delicate the dance of science and policy can be, and deserves a closer look.

Publishing a scientific paper is in some ways like sending a painting off to a gallery, or a book to a publisher. You take a deep breath, and wait for the critics to respond. When reviewing papers for publication in science journals, the part of a paper I looked most closely at – but which receives almost no interest in the popular press – is the *Methods* section. The popular press is always ready to jump to the *Conclusions*. But the results and conclusions are only as reliable as the methods, which must be carefully-designed and meticulously carried out. As soon as it appeared, the Losey study went under the microscope. Others – yes, some of whom as employees of the chemical industry had a conflict of interest – found problems with some of the methods used in the research. Here's what they concluded.

When genetically engineering a plant, researchers insert not only the gene they want in that species, but a piece of DNA which regulates the expression of that gene. That segment of DNA, called a promoter, is like a volume dial on a stereo. It regulates how much protein is made from the gene. There's a whole suite of promoters biotechnologists can choose from. With the *Bt* corn, depending on the kind of promoter attached to the protein gene, different levels of the Bt- protein show up in different strains of the corn, and also in different tissues. One kind of *Bt* corn ("event-176") was engineered to produce the protein in its pollen at levels 40 or more times as high as in other tissues, compared to that normally expressed by most *Bt*- engineered corn. This appears to be the strain used by Losey *et al*. Event-176 corn is no longer available to farmers, nor are its very high levels of *Bt* protein in its pollen. Secondly, the levels of pollen Losey *et al* exposed their monarch larvae seemed to be considerably higher than levels monarchs would likely be exposed to in real life. These two issues with the study's design led many to question the conclusion that the use of *Bt*- engineered corn could seriously affect monarch populations.

Wisniewski et al [26] went on to argue that no one has been able to replicate Losey *et al*'s original work without using pollen densities on milkweed leaves well above those ever found in nature. Other work indicated that though the event-176 strain did kill monarch larvae in field studies, other strains in which the *Bt* was not so highly expressed in the pollen had no discernable effect.[27]

One study showed that lepidopteran abundance and diversity in weeds near *Bt* corn was no different from untreated corn fields, but in weeds near fields treated with chemical herbicides both abundance and diversity were reduced.[28] This suggested that in the choice between chemical herbicides and *Bt* corn, the latter was less harmful to non-target insects.

The question of how much *Bt* corn might affect monarchs is also an example of how "timing is everything." When *Bt*-corn plants produce their pollen is critical to their toxicity. If there are no monarch larvae in or near cornfields when pollen is being produced, it doesn't matter how toxic the pollen is. So, if exposure to it is zero, harmful effects should be zero. Oberhauser *et al* [29] concluded that at least in Iowa corn and soybean fields produced overall almost 80 times as many monarchs as non-agricultural areas. If you drive through Iowa it's clear why this is – almost the entire landscape is devoted to corn and soybeans, with little non-agricultural land left. Similar estimates were made for Minnesota and Wisconsin. In Iowa it appeared that over the course of a year about 3% of monarch larvae might be exposed to pollen from *Bt*-engineered plants. Exposure in more southerly parts of the monarch breeding range, where corn and soybeans are less dominant, would be expected to be significantly less. So any harmful effects caused by exposure to the *Bt* pollen – say a 10% reduction – would be only in those 3% likely to be exposed to the pollen. To those monarchs experiencing the harm, this is a real effect, but overall unlikely to be significant to the entire monarch population. Angharad *et al* [30] looked at a number of studies of the effect of *Bt* corn on monarchs and other beneficial insects and concluded there was little evidence the engineered corn could, or would, affect these populations.

But applying the precautionary principle – better safe than sorry – leaves some questions still unanswered. Research showed that early-instar monarch larvae can be affected by being exposed to the anthers or pollen of *Bt*-corn without actually eating them. A 21% decrease in weight gain in these larvae was noted in those larvae who were placed on milkweed leaves with Bt-corn anthers compared to those placed on leaves with non-engineered corn anthers.[31] Karen Oberhauser is one of the most active and respected scientists studying monarch biology. Her and Michelle Solensky's website Monarch Butterfly Ecology http://www.ecology.info/monarch-butterfly.htm cites research which indicates that pollen naturally deposited on milkweed plants in corn fields does affect monarch survivorship. She concludes that "This finding suggests that the blanket conclusion that *Bt* corn poses no risks to monarchs…should be revisited."

So what do we do with the precautionary principle? Experience suggests that when lots of money can be made, e.g. by the creators of the genetically modified plants, or saved – e.g. by the farmers using them – it's easy to slip the precautionary principle into a drawer and out of sight. That's especially true when the potential harm is caused to some other species than our own. But those of us who have lived through the near-extinction of our national symbol, the bald eagle, in the lower-48 states and seen it rebound because of the banning of DDT, know how useful the precautionary principle can be. Unfortunately in the case of DDT it took decades to make that clear. In the fall of 2007 European Union environmental officials proposed banning several kinds of genetically-modified corn, in part because of concern over their effects on monarchs. In Europe, however, there is much more distrust of genetically modified crops in general than in the U.S.

The use of herbicides and herbicide-resistant genetically-modified plants may have other effects on monarch populations. On the Monarch Watch website[32] Chip Taylor points out that the use of *Roundup-Ready* soybeans, by allowing farmers to apply higher levels of the herbicide *Roundup* to their crop reduces "weed" populations, like milkweeds, within the fields. In the last ten years or so *Round-up Ready* soybeans may have caused the loss of some 80 million acres of monarch habitat. This may already be causing a shift in monarch populations from being centered in the corn-belt (and increasingly soybean-belt) of the Midwest, to farther east where they reproduce along roadsides and in ditches. The recent emphasis on corn ethanol to replace foreign petroleum may also change the feeding/breeding landscape for monarchs, as will the conversion of fallow fields to switchgrass or other biofuels. Already in 2007 some 92 million acres of *Roundup-Ready* corn was planted, about 14 million acres more than in 2006. Chip Taylor's estimate is that more than 100 million acres of milkweed have been lost due to the introduction of Round-up Ready crop varieties.[33]

In and around cities and industrial areas, levels of the pollutant ozone can rise to significant levels. Milkweed species have been found to be particularly sensitive to ozone.[34] Dwindling milkweed populations because of this – and perhaps other as yet unidentified sources of pollution – could affect monarch populations.

A detailed survey of population trends of 40 different kinds of butterflies across 820 sites in England, carried out by the group Butterfly Conservation, showed that butterflies in agricultural areas had declined by 30% in the last

ten years. A chilling article in the *Guardian* newspaper documents the serious loss of butterflies in England.[35] One study of 46 species of butterflies there indicated that three-fourths were on the decline due to habitat loss.[36] On the continent the Maderian Large White has recently gone extinct and the Grecian copper is in danger of the same. The status of butterfly populations in six countries is worse than in Britain.[37] Twenty-three species of butterflies have been designated as endangered or threatened in the U.S. – fifteen of them in California! A disturbing but well-done article on threats to U.S. butterflies can be found at:
http://www.nwf.org/nationalwildlife/article.cfm?issueID=122&articleID=1596

Monarchs are only one butterfly species of a hundred or more found in most American states. They're the large charismatic species that almost everyone recognizes, and have been called "the most ecologically-researched non-pest arthropod."[38] A few species of butterflies have evolved to eat pollen as adults. We have essentially no idea what the effects of *Bt*-corn or other agricultural practices are on these other species, equally deserving of protection.

Our impacts on monarchs come from more than just cutting down the fir forests of Mexico or using new technologies on our crops. We kill them, one by one, with our vehicles. One scientist counted butterflies killed along highways in central Illinois, and estimated as many as 20 million were killed each summer by vehicles in Illinois alone. Of these perhaps 500,000 were monarchs.[39] The math is easy. If this number is anywhere near right, the inevitable conclusion is that somewhere in the tens of millions of monarchs are killed every year in traffic fatalities.

The ditches and right-of-ways along our highways often have milkweed growing in them, which of course attracts the butterflies. As I write this our township mower is passing by, mowing grass and weeds along the road past my house. Some of the milkweed plants on which I've found eggs and larvae now lie drying in the swath. Unless the larvae are large enough to find and move to a live milkweed, they will starve. The mowing of these areas must each summer in the U.S. result in millions of monarch deaths. In other places weeds along roads are sprayed with herbicides, also toxic to non-target species. The nascent movement toward more natural, un-mowed grasses along highways is clearly a move in the right direction.

Then there's all the alien species we humans have introduced to monarch habitats. Though larvae will eat any of the hundred or so species of milkweed found in the U.S., the most important food for larvae growing up in the east and midwest is *Asclepias syriaca*. Black swallowwort (*Cynanchum nigrum*, also called *Vincetoxicum nigrum*), has made its way into the eastern U.S. as an invasive plant from Europe.[40] It was first reported from Massachusetts, in 1854. Though a member of the milkweed family on which female monarchs will sometimes oviposit, because it is not a good source of food for the larvae, few survive on it.[41]

Alien species are often strong competitors to native species. Dog-strangler vine and several species of the plant genus *Vinetoxicum* are introduced species which inhabit similar habitats as milkweed, which in some cases they replace. Female monarchs ready to oviposit tend to shun these plants and unless they can find milkweed may not lay eggs at all. It's a sensible choice, because larvae given only these species to eat don't survive.[42,43] I've watched roadsides in northern Wisconsin being taken over by spotted knapweed, an aggressive alien which though nectared on by monarchs, shoulders out native wildflowers, including milkweed, goldenrod and other nectar sources. Some ditches and their nearby hillsides have become monoculture fields of knapweed.

Fire ants, an especially aggressive species of ant, arrived in Mobile, Alabama from Brazil during the 1920s, and have since spread throughout much of the southern states. In some parts of Texas mounds are as dense as 1000 mounds per acre. They are voracious predators which will attack and kill rats. A number of insect species have declined as the ants became more populous. Fire ants have been seen attacking and consuming monarch larvae. They may be an important factor in the distribution of monarchs, especially affecting the first generation of spring migrants. At present fire ants are found in east but not west Texas. Lepidopteran densities in the Austin area of east Texas appear to be less than half what they were prior to the invasion of the fire ant.[44]

A new species of wasp which came to the U.S. around 1980, the European paper wasp (*Polistes dominulus*) feeds its young small caterpillars. There is already some evidence in the Denver, Colorado area that this wasp has decreased butterfly populations.[45] It's too early to tell what its effect on monarchs has been or will be.

Another recent invasive, brought into the U.S. as a predator of undesirable insects such as aphids in orchards, is the multicolored Asian lady beetle *Harmonia axyridis*. Those of us who've had to deal with their migration into our houses every fall are only too intimate with this pest. High populations are found in soybean fields, where they prey on the soybean aphid. Unfortunately they also prey on early instar monarch larvae. The spread of soybean cultivation has helped this species immensely – one way in which grain future trading on global markets reaches down into local ecology. Asian lady beetles appear not to have much effect on monarch larvae growing up on milkweeds in corn fields, where few lady beetles are found. But there is a moderate to high risk on the monarch larvae born in or near soybean fields.[46] Fortunately milkweed is much less common in soybeans than in corn. And it appears the lady beetles prefer aphids to monarch larvae — the monarch's "armor" of chemical defense seems to be working. So where aphids are present in soybeans, monarchs are even less affected.[47]

Our species' manipulation of the natural environment has cast a shadow over many other species' populations. One study found that fire-suppression in the Ouachita Mountains of Arkansas, a region through which many southward-migrating monarchs pass, has decreased the density of plants the monarchs rely on for nectar.[48]

I spent nearly twenty years inside a classroom teaching biology. Over those years it became blatantly obvious that fewer and fewer college-aged students knew anything to speak of about nature. Most could not identify a meadowlark or heron, or the names of the commonest wildflowers in the woods, fields and ditches around them. I demonstrated their ignorance to them by giving a quiz. I showed five pictures of pop-media icons (rock stars, movie stars, athletes.) Ninety to ninety-five percent of them would correctly identify the picture. I'd follow that with a picture of a common bird, mammal, or insect. Generally fifteen to twenty percent, sometimes much less, would get that right. Monarch adults, commonly identified, were the exception. We then discussed why they knew the media icons but not their wild neighbors all around them.

Few of my students, with the exception of those who hunted and/or fished, spent time outside in nature. They suffer a syndrome called "nature deficit disorder." Thousands of generations of our ancestors walked through nature, and even more recently as farmers spent their days in it. Those of us who do hike the woods and fields know the spiritual/emotional benefits. Those who don't suffer; and other species suffer, too, because you can't really care about what you don't know. Recently (2007, and reintroduced again in 2009) legislation to address this has been brought to the U.S. Congress in the form of the No Child Left Inside Bill. This and any other efforts at getting students of all ages out into nature need your support.

All these large forces – our highways, pesticide technologies, the geopolitics of logging and agriculture, invasive species, nature deficit disorder – whittle away at monarch populations. (For a well-written overview of the struggles of butterflies in a global economy, see http://www.orionmagazine.org/index.php/articles/article/546/).

Individually you and I can send out letters to editors, contact local or national politicians, and like myself pull up alien plants when we find them. This seems sometimes a Sisyphean effort. A more pleasurable and equally useful work is to seed your garden or lawn with plants that attract butterflies. Your local county extension agent or salesperson at a nursery can help you choose annuals and perennials which attract butterflies. Migrating monarchs need service stations in which to refill themselves for the continued flight, and as discussed earlier, for enough lipids to survive the long winter. Literally thousands of these refueling stops for migrating monarchs have popped up around the U.S. and Canada, thanks to efforts by Monarch Watch's Monarch Waystation program. These Waystations are small oases in an increasingly unfriendly landscape for the migrating monarchs. These sites can be created in your garden, on abandoned or conservation reserve land, along roadsides, or especially on school property where students can observe, enjoy, and learn to appreciate monarchs.

See the *Actions* section of this book for other suggestions on how to help the monarch.

Epilogue:
Emergencies

By devouring and digesting the earth's offerings we sustain ourselves. With our imaginations we transmute these same offerings, making of them our lives. We are not a species – none are – that can live in a vacuum. Our molecules are the earth's, were once other creatures'. They pass through us. We use and transform them, second by second.

Social and intellectual creatures that we are – as other creatures are, to their own extent and ends – we absorb too the ideological molecules which our traditions have synthesized over the centuries and millennia. Chemical molecules, I think, are more easily metabolized than these building blocks of our own consciousness and ethics, which Richard Dawkins has called "memes."

Like ruminant larvae we ingest defensive beliefs to protect ourselves. Deep-seated beliefs about who we are in relation to other creatures: We are the crown of creation. Only we have immortal souls. The earth and its creatures have been laid out before us as a great banquet for us to feast on.

Through protecting ourselves and our habits of consumption with these toxic beliefs has come an emergency. It is nigh used-up, the table of our delights, and a reckoning is due. Other kinds we have made extinct, and not just a few. To them we were a sudden storm, and many now are only left to look out over the ruins of their ancient but once vital village. Air, earth and water carry the signature of our consumption, credit overdue.

Like late-instar larvae, we are hulking, consuming machines. The urge to transform whispers gently in our inner ears. The time has come to pause, in-crypt ourselves, enter the dangerous ecstasy of transmutation, and emerge.

Lightly then to tread the earth, and wing our way in migrations to new landscapes of belief, huddling and sharing and looking out for others, of all kinds.

Resources

The many resources listed below are not meant to be complete.
My apologies to those I've not included.

Organizations

Monarch Watch: Arguably the premier web resource for monarch enthusiasts, at http://www.monarchwatch.org. A repository of great interest and value, with many links to other sites. Contains educational, research, and monarch conservation resources. Our thanks to Chip Taylor for setting this site up and keeping it up. Additional resources and websites, not shown below, are linked to the Monarch Watch site, including the Monarch Watch Blog: http://monarchwatch.org/blog/.

The Monarch Butterfly Sanctuary Foundation: A U.S. group which is active in support of conserving the monarch Reserves in Mexico. http://www.mbsf.org/

Monarch Larva Monitoring Project (and MonarchLab): An NSF-funded citizen-science project (thanks to Karen Oberhauser) to understand monarch larva ecology and disseminate the results of the research. Volunteers do weekly surveys at many sites in the eastern U.S. during the summer breeding season; their and other reports are archived here, along with lots of other resources. http://www.mlmp.org. (Also see Karen Oberhauser's really comprehensive and attractive Monarchlab Website http://www.monarchlab.umn.edu/).

Live Monarch Foundation: A nonprofit educational & conservation foundation (with an emphasis on preserving monarch habitat.) http://www.livemonarch.org/

Monarch Buttterfly New Zealand Trust: http://monarch.org.nz/

Monarchbutterfly.org: Created and funded by the Pismo Beach Visitors Bureau, a website which highlights and honors the monarch and especially Pismo Beach, CA where California's largest wintering site is found. http://www.monarchbutterfly.org/

The Association for Butterflies: http://www.forbutterflies.org

Cape May, N.J. roadside census reports, going back to 1992: http://rkwalton.com/mmp02.html

The Butterfly Website: http://butterflywebsite.com/

Xerces Society: Dedicated to invertebrates, including but certainly not limited to insects. An especially relevant project is: The First Butterfly Big Year. During the year 2008 Xerxes-founder Robert M. Pyle traveled the U.S. and Canada to find as many species of butterflies as possible. He kept a blog, and his results will be published in a book *Swallowtail Seasons: The First Butterfly Big Year*. For more information see: http://www.xerces.org/Butterfly_Conservation/butterflyathon.html

Project Monarch Alert: Tagging, studying and interpreting the migratory habits of monarchs west of the Rockies: http://www.calpoly.edu/~bio/Monarchs/

Alternare: A Mexican organization dedicated to preservation of natural resources and the rural community way of life (website in Spanish.) http://www.alternare.org/sitio/. Donations in American dollars can be made to Alternare through: http://www.mbsf.org/

Ecolife Foundation: Their website describes their efforts to help the Mexicans living in and near the monarch Reserves: http://ecolifefoundation.typepad.com/

Southwest Monarchs study: Studying the migration of monarchs through the southwest U.S. http://swmonarchs.org/index.php

Tours of the over-wintering sites

http://www.spiritofbutterflies.com/butterflytour.html

Through the Univ. of Florida: http://www.flmnh.ufl.edu/butterflies/expeditions.htm

Or contact the Monarch Teacher's Network:
http://www.eirc.org/website/Programs-+and+-Services/Monarch-
Teacher-Network.html

Monarch Festivals

A number of states and municipalities sponsor butterfly or specifically monarch festivals. Some have adopted the monarch as their official symbol. Examples include:

The Texas Butterfly Festival, each October, sponsored by the Greater Mission, Texas, Chamber of Commerce.
In Grapevine, Texas, near the end of October, the annual Butterfly Flutterby.
St. Mark's National Wildlife Refuge, NE Florida, middle of October.

Books and Movies

(In addition to literature cited in the chapters above. Note – this lists does not include a very extensive juvenile literature concerning the monarch, and is not meant to be complete.)

Milkweed, Monarchs and More: A Field Guide to the Invertebrate Community in the Milkweed Patch — A 96-page, full color field guide to the invertebrates found in milkweed written by Ba Rea, Karen Oberhauser and Michael A. Quinn. Available from: http://www.monarchlab.org/store/pc-18-2-milkweed-monarchs-more.aspx

Handbook for Butterfly Watchers, by Robert Michael Pyle, 1992.

The Monarch Butterfly and *The Monarch Butterfly: International Traveler,* by Fred Urquhart.

Chasing Monarchs by Robert Michael Pyle.

Four Wings and a Prayer by Sue Halpern, and the award-winning movie *Four Wings and a Prayer*.

An Obsession with Butterflies: Our Long Love Affair with a Singular Insect, by Sharman Apt Russell, 2003.

The Monarch Butterfly: Uniting a Continent (available in Spanish and English)

Monarch Butterflies: saving the King of the New World and *The Last Monarch Butterfly: Conserving the Monarch Butterfly in a Brave New World*, by Phil Schappert.

Las mariposas entre los antiguos Mexicanos, by Carlos Beutelspacher, a detailed cultural history of butterflies in Mexico, in Spanish, 1988.

On the Highway of Butterflies, a German film featuring Robert Michael Pyle.

Papalotzin: The Flight of the Monarch DVD, available from Nieman-Marcus.

The Butterfly Trees, a one-hour documentary by Kay Milam, in progress.

Monarch Butterfly Biosphere Reserve, from Lighthawk Reconnaissance, an 11-minute DVD of the Biosphere Reserves, available from Dr. Lincoln Brower, SweetBriar College, Va. 22495. $20 by check includes shipping.

Don't: The Metamorphosis of a Monarch Butterfly, Phoenix Learning Group, Inc. A DVD- movie made from the point of view of a monarch butterfly, "...a study of the lyric passage of a Monarch butterfly from its birth to its delicate metamorphosis from caterpillar into butterfly, and also of its journey from country to city..."

The Monarch: A Butterfly Beyond Borders, Bullfrog Films.

"Nomads of the Wind: The Journey of the Monarch Butterfly and other Wonders of the Butterfly World", by Ingo Arndt, 2009. A lovely photo-essay

following the migration of monarchs from the U.S. midwest to Michoacan, Mexico.

The Incredible Journey of the Butterflies, a PBS movie which aired on January 27, 2009. Available from http://www.pbs.org. Transcript free of charge at http://www.pbs.org/wgbh/nova/transcripts/3601_butterflies.html

Various Other Websites

The website of Dr. Anurag Agrawal at Cornell Univeristy has a nice monarch biology link: http://www.eeb.cornell.edu/agrawal/index.html

Texas Monarch Watch: http://www.texasento.net/dplex.htm

Monarch parasites: http://www.monarchparasites.org/

Tagging California monarchs: http://www.monarchprogram.org/tagging.htm

Distribution, Status and Conservation of the monarch in Canada: http://web.biosci.utexas.edu/philjs/Monarch/monrep97.html#Population

"Flight of the Monarchs," from Vanity Fair, Nov. 1999 (with an especially detailed account of the discovery of the over-wintering sites): http://www.dispatchesfromthevanishingworld.com/pastdispatches/monarch/pr intermonarch.html

The journey of the Papalotzin, a lightweight airplane, following the monarchs (in Spanish): http://www.papalotzin.com/esp/ and http://www.wwfca.org/php/news/reportajes/Agosto05.php

The uneasy relationship between the Mexican peasant and the monarchs: http://www.planeta.com/planeta/98/0598monarch.html

Lots of nice photos of various stages of the monarch, including one of an egg hatching:
http://picasaweb.google.com/jim.c.ellis/Monarchs2007EggHatchingPupa FormingSexDiifferentiationInPupae

A database of collection sites for Florida's Museum of Natural History (Gainesville) collection of more than 3 million specimens of moths and butterflies, a kind of map of lepidopteran species of North America, linked at the Butterflies and Moths of North America website: http://www.butterfliesandmoths.org.

The first seven minutes or so of this film about migrations is about the monarch migration: http://www.hulu.com/watch/63304/amazing-journeys

A video detailing the work of the Alternare community organization helping farmers in and near the monarch reserves: http://monarch.pwnet.org/mom/cim_webcast.php

A selected bibliography of monarch butterfly references: http://www.loc.gov/rr/scitech/SciRefGuides/butterfly.html

Butterfly gardening

Butterfly Gardening: Creating Summer Magic in Your Garden, from the Xerces Society.

Attracting Hummingbirds and Butterflies to Your Backyard, by Sally Roth.

http://www.thebutterflysite.com/gardening

http://www.monarchwatch.org/garden/index.htm

Janet Allen's Stewardship Gardening website (creating gardens for wildlife, including the monarch): http://www.stewardshipgarden.org/creatures/monarch.html

Also see: http://wildlifefriendlylawn.blogspot.com/

Link to an article from Mother Earth News about attracting butterflies: http://www.motherearthnews.com/Nature-Community/2004-06-01/Bring-in-Butterflies.

Using monarchs in the classroom

MonarchLive: A classroom distance-learning adventure following the monarch migration. http://monarch.pwnet.org/

The Monarch Teacher Training Network has many useful ideas for how to use monarchs in math, science and other classes at various levels::
http://www.eirc.org/website/Programs-+and+-Services/Monarch-Teacher-Network.html, including a 30- minute DVD called "Journeys and Transformations" useful in the classroom.

Journey North – Monarch Migration for Schoolchildren. Includes links to classroom lessons. One project is called the "Symbolic Migration." Students make their own paper butterflies and send them, along with donations to help support monarch conservation, to other students in Mexico. In the fall of 2007 nearly 1500 schools were involved. The website is full of useful information and curriculum suggestions. From it students can compare the over-wintering population numbers over the years or track the northward migration of the butterflies in the spring. http://www.learner.org/jnorth/monarch/

Monarch Teacher Network-Canada: http://www.monarchcanada.org/

Monarch Lab.Org: At http://www.monarchlab.org, already mentioned, many resources for classroom teachers who use monarchs in their curriculum. I've used this site regularly. Includes a link to a new Schoolyard Ecology grant program. For curriculum guides: http://www.monarchlab.umn.edu/store/c-2-books-curriculum-guides.aspx

Live Monarch —They have an adopt a larva program in which children are sent regular emails on the progress of their caterpillar — http://www.livemonarch.com/adopt.htm

Take a Child Outside: http://www.takeachildoutside.org/

The Children & Nature Network, http://www.cnaturenet.org

An article about efforts to conserve the monarchs, esp. among schoolchildren — *Safe Haven*, Emily Costello. Science World. New York: Mar 10, 2008, Vol. 64:11, p.8.

Workshops targeting children and supporting the Monarch Butterfly Sanctuary Foundation:
http://www.communityartsadvocates.org/monarchbutterfly.html

Monarchs and Milkweeds, a DVD targeting the elementary classroom:
http://www.nutmegmedia.net/MAIN/Dist_List.htm

Commercial Butterfly Ventures

http://www.butterflyab.com/

http://www.greathousebutterflyfarm.com/

http://www.kirkwoodbutterfly.com/

http://www.allaflutterbutterflies.com/

http://www.vibrantwings.com/

http://www.butterflyrick.com/

http://www.madambutterfly.co.nz/

Actions

A few suggestions of what YOU can do to help the monarchs, who struggle not only against predators, parasites, the weather, and diseases but against the forces of human activities.

WALK WITH YOUR FAMILY AND FRIENDS AND SHOW THEM MONARCHS AND THEIR WAY OF LIFE. We only care about what we know about, and it's not possible to know these sweet, incredible creatures without becoming somehow attached to them.

LEARN MORE ABOUT MONARCHS YOURSELF. The more you know, the more you'll care...and be able to share with others.

MAKE A SHORT (OR LONG) PRESENTATION ABOUT MONARCHS at your favorite club, church organization, etc. Let others know about these fascinating creatures...

PLANT A MONARCH WAYSTATION. See especially www.monarchwatch.org/waystations

TRAVEL TO MEXICO TO SEE THE INCREDIBLE OVER-WINTERING CLUSTERS: While you're there, tell everyone you can you've come because of the monarchs.

SUPPORT THE EDUCATIONAL OR PRESERVATIONIST ORGANIZATION OF YOUR CHOICE.

Index

Chapter Notes

Chapter 1 Notes

[1] *Science*, January 14 (2000)

[2] Folsom, E. Virginia Quarterly Review, Spring (2005)

[3] Malcolm *et al* in *North American Conference on the Monarch Butterfly* (1999)

[4] (http://www.monarchwatch.org/update/2006/0831.html#4)

[5] Don Wilks, Dplex-l listserv.

[6] Barb Case, Dplex-l listserv

[7] Ent. Exp. et Appl. 19(2), (1976)

[8] Densmore, *How Indians Used Wild Plants for Food, Medicine and Crafts*, Dover.

[9] Foster & Duke, *Peterson Field Guides: Eastern/Central Medicinal Plants*, Houghton Mifflin, 1990

[10] Entom. Exp.et Appl. 104(2), (2002)

[11] Harlen and Altus Aschen, Dplex-l listserv.

[12] Ecol. Entomol. 26(2), (2001)

[13] Zalucki *et al*, Aust. Ecol. 26(5), (2001)

[14] Malcolm, Cockrell and Brower, J. Chem. Ecol. 15(3), (1989)

[15] Nelson, in *Biology and Conservation of the Monarch Butterfly* (1993)

[16] Chemoecol. 3(2), (1992)

[17] Bob Morton, Dplex-l listserv

[18] http://www.mlmp.org/

[19] Holzinger, Frick and Wink, FEBS Lett. 314(3), (1992)

[20] Labeyrie and Dobler, Mol. Biol. Evol. 21(2), (2004)

[21] Bethann and Oberhauser, Env. Entom. 33(4), (2004)

[22] J. Chem. Ecol. 21(5), (1995)

[23] Rothschild and Marsh, Ent. Exp. et Appl. 24(3), (1978)

[24] Oecol. 42(3), (1979)

[25] Fordyce, Marion and Shapiro, Journal of Chemical Ecology 31(12), (2005)

[26] Solensky and Larkin, Ann. Ent. Soc. Am. 96(3), (2003)

[27] Zalucki, Clarke and Malcolm, Ann. Rev. Entomol. (47), (2002)

[28] http://www.monarchwatch.org/biology/sexing.htm

[29] Nijhout and Emlen, PNAS (95), (1998)

[30] Nijhout and Grunert, PNAS (99), (2002)

[31] Brakefield and French, BioEssays (21), (1999)

[32] Pelling, *et al*, J. Royal Soc. Interface, August 5 (2008)

[33] Gimzewski, J.K. Journal of the Royal Society *Interface* January 6, *(*2009)

[34] Liu, S-S. and Liu, T-X, Ecol. Entom. (31), (2006)

[35] Weiss, M. New Scientist, March 5 (2008)

[36] Yack and Fullard, Nature (403) (2000)

[37] Kinoshita, M., Shimada, No. and Arikawa, K., Journ. Exper. Biol. (202), (1999)

[38] Weiss, M., Anim. Behav. (53), (1997)

[39] Weiss, M. and Papaj, D., Anim. Behav. (65), (2003.)

[40] Anim. Beh. (31), (1984)

[41] Dukas, R. and Bernays E.A., PNAS (97), (2000)

[42] Nuttman, C. and Willmer, P., Ecol. Entom. (28), (2003)

[43] Schappert, Nat. Hist. July (2001)

[44] Masters, in *Biology and Conservation of the Monarch Butterfly* (1993)

[45] Kammer, J. Comp. Phys. 68(3), (1970)

[46] http://www.monarchwatch.org/update/2006/0930.html#2

[47] Herman *et al*, J. Expt. Zool. 218(3), (2005)

[48] Barker and Herman, J. Exptl. Zool. 183(1), (2005)

[49] Herman and Tatar, Proc. Biol. Sci. 268(1485), (2001)

[50] Goering and Oberhauser in *North American Conference on the Monarch Butterfly* (1999)

[51] Goehring and Oberhauser, Ecol. Entoml. 27(6), (2002)

[52] Pavel Friedman, *The Butterfly,* from *I Never Saw Another Butterfly*, Hana Volavkova, ed. (1993)

Chapter 2 Notes

[53] Chaves, *et al* J. Insect Behav. 19(2), (2006)

[54] Berenbaum, in *Insect Lives*, ed. by Hoyt & Schultz, (1999)

[55] Bull, *et al*, Aus. Ecol. 10(4), (1985)

[56] Rutowski, Behav. Ecol. and Sociobiol. 7(2), (1980)

[57] Ba Rea, Dplex-1 listserv.

[58] Zalucki, in *Biology and Conservation of the Monarch Butterfly* (1993)

[59] Zalucki and Kitching, Oecol. 53(2), (1982)

[60] Wedell, J. of Exp. Biol. 208, (2005)

[61] Nystrom and Backvall, J. Org. Chem. 48(22), (1983)

[62] Schneider, in *Biology and Conservation of the Monarch Butterfly* (1993)

[63] Solensky, J. Insect Behav. 17(6), (2004)
[64] Svard and Wiklund, Oikos 51(3), (1988)
[65] Karlsson, Leimar and Wiklund, Proc. Royal Soc. B 264 (1381), (1997)
[66] Svard and Wiklund, Behav. Ecol. Sociobiol. 23(1), (1988)
[67] Karlsson, Ecol. 79(8), (1998)
[68] Oberhauser, Anim. Behav. 36(5), (1988)
[69] Rogers and Wells, J. Morph. 180(3), (1984)
[70] Oberhauser, Behav. Ecol. Sociobiol. 25(4), (1989)
[71] Oberhauser, Funct. Ecol. 11(2), (1997)
[72] Bissoondath and Wiklund, Behav. Ecol. Sociobiol. 39(5), (1996)
[73] Svard and Wiklund, Behav. Ecol. Sociobiol. 24(6), (1989)
[74] Oberhauser, Behav. Ecol. Sociobiol. 31(5), (1992)
[75] Forsberg and Wiklund, Behav. Ecol. Sociobiol. 25(5), (1989)
[76] Kaitala and Wiklund, Behav. Ecol. Sociobiol. 35(6), (1994)
[77] Kaitala and Wiklund, J. Insect Behav. 8(3), (1994)
[78] Oberhauser and Hampton, J. Insect Behav. 8(5), (1995)
[79] Bissoondath and Wiklund, Behav. Ecol. Sociobiol. 37(6), (1995)
[80] Solensky, M.J. and K.S. Oberhauser, Animal Behaviour, 77 (2), February, (2009)
[81]

http://newsgroups.derkeiler.com/Archive/Sci/sci.bio.entomology.lepidoptera/2009-02/msg00016.html
[82] Frey, *et al*, J. Lepid. Soc. 52(1), (1998)
[83] VanHook, in Biology and Conservation of the Monarch Butterfly (1993)
[84] Brower, L. *et al* J. Lepidopterists Society 62(4), (2008)
[85] Wells, Wells and Rogers, in *Biology and Conservation of the Monarch Butterfly* (1993)
[86] Oberhauser and Frey, in *North Amercian Conference on the Monarch Butterfly* (1999)
[87] Frey, D. in *North American Conference on the Monarch Butterfly* (1999)
[88] Baur *et al*, Phys. Ent. 23(1), (1998)
[89] Ladner and Altizer, Ent. Exp. et Appl. 116(1), (2005)
[90] Herman, in *Biology and Conservation of the Monarch Butterfly* (1993)

Chapter 3 Notes

[1] Theis *et al* , Ann. Missouri Bot. Garden 94(2), 2007
[2] Farrell and Mitter, Biol. J. Linn. Soc. 63, (1998)

[3] Smith and Allen, Zool. J. Linnean Soc. 144(2), (2005)
[4] Berenbaum, Am. Midl. Nat. 111(1), (1984)
[5] Baldwin, *et al*, Science 311, (2006)

[6] Anurag A. Agrawal and Mark Fishbein (2008), *Proccedings of the National Academy of Sciences 105:10057-10060.*

[7] Van Zandt and Agrawal, Ecology 85(9), (2004)
[8] Brower *et al*, J. Chem. Ecol. 8(3), (1982)
[9] Lynch and Martin, in *Biology and Conservation of the Monarch Butterfly* (1993)
[10] Rothschild, Phytochem. 45(6),(1997)
[11] Rothschild, Moore and Brown, Biol. J. Linn. Soc. 23(4),(1984)
[12] Boppre, *The Monarch Butterfly: Research and Conservation* (1993)
[13] Chai and Szygley, Am. Nat. 135(6), (1990)
[14] Marden and Chai, Am. Nat. 138(1), (1991)
[15] Brower and Glazier, Science 188 (4183), (1975)
[16] Arellano *et al*, in *Biology and Conservation of the Monarch Butterfly* (1993)
[17] Glendinning, in *Biology and Conservation of the Monarch Butterfly* (1993)
[18] Roode, *et al,* Journal of Animal Ecology, 77(1), (2008)
[19] Biotropica 20(1) (1988)
[20] Brower, Evol. 12(1), (1958)
[21] Ritland, Oecologia 88(1), (1991)
[22] Prudic, *et al*, Journal of Chemical Ecology, 33(6), (2007)
[23] Ritland and Brower, in *Biology and Conservation of the Monarch Butterfly* (1993)
[24] Ritland, *Ecology* 75(3), (1994)
[25] Alonso-Mejia and Brower, Cell. Mol. Life Sci. (50)(2), (1994)
[26] Angrawal, A. Natural History, Sept. (2002)
[27] Stimson and Kasya, J. Lepid. Soc. 54(1), (2000)
[28] Jiggins *et al*, Parasit. 120(5), (2000)

Chapter 4 Notes

[29] http://www.time.com/time/magazine/article/0,9171,867518,00.html
[30] Brower, L. P. "Understanding and misunderstanding the migration of the monarch butterfly (Nymphalidae) in North America 1857-1995." *Journal of the Lepidopterists' Society*, v. 49, December, (1995)
[31] Dockx *et al*, Eco. Appl. 14(4), (2004)

[32] Dockx, C. Biol. J. Linn. Soc. 92(4), (2007)

[33] Williams, C.B. *Insect Migration*, Macmillan (1958)

[34] May, Michael L. and J.H. Matthews, in Drgaonflies and Damselflies, August (2008)

[35] Zhu, H. *et al* Current Biol. 15, (2005)

[36] Reppert, S., Cell, 124(2), (2006), online at http://biology.plosjournals.org/archive/1545-7885/6/1/pdf/10.1371_journal.pbio.0060004-L.pdf

[37]

http://online.wsj.com/article/SB120233944314148959.html?mod=googlenews_wsj

[38] Hesman, T. Science News, April 25, (2009)

[39] Merlin, C., Gegear, R. and Reppert, S. Science Scept. 25, (2009)

[40] Labhart and Meyer, Curr. Opinions in Neurobiol. 12, (2002)

[41] Reppert, Zhu and White Curr. Biol. 14(2), 2004

[42] Sauman *et al*, Neuron 46(3), (2005)

[43] Perez, S., O. R. Taylor, and R. Jander. A Nature 387: 29 (1997)

[44] Perez, Taylor, and Jander, in *North American Conference on the Monarch Butterfly,* (1999)

[45] Oliveira, Srygley and Dudley, J. Exp. Bio. 201(24), (1998)

[46] Lohmann, Nature, April, (2004)

[47] Heyers, *et al*, Public Library of Science Oct. 1, (2007)

[48] (Etheredge *et al*, PNAS 96(24), (1999)

[49] Jones, D.S. and MacFadden, B.J., Journ. Exptl. Biol. (96), (1982)

[50] Srygley, R.B. *et al*, Anim. Beh. 71, (2006)

[51] Phil. Trans. R. Soc. Lond. B, 337, (1992)

[52] Dingle *et al*, Biol. Jr. Linn. Soc. 85(4), (2005)

[53] Calvert *et al*, J. Lepid. Soc. 55(4), (2001) (available at http://research.yale.edu/peabody/jls/pdfs/2000s/2001/2001-55(4)162-Calvert.pdf

[54] Helbig, A.J. Journal of Experimental Biology 199, (1996)

[55] Biro *et al*, PNAS 104, (2007)

[56] Able and Able, Letters to Nature, 347, (1990)

[57] Schmidt-Koenig, in *Biology and Conservation of the Monarch Butterfly*, (1993)

[58] Walker, T.J. in *Migration: Mechanisms and Adaptive Significance*, (1985)

[59] Chip Taylor, Dplex-l discussion list, Nov. 1, (2007)

[60] Rio Bravo Nature Center, on Dplex-l list

[61] http://www.leeric.lsu.edu/le/cover/lead094.htm

[62] Srinivasan *et al*, Science 287, (2000)

[63] Canovaso *et al,* "Fat Metabolism in Insects," Annu. Rev. Nutr., (2001)

[64] Seiber *et al*, J.Chem. Ecol. 12(5), (1986)

[65] Brower, Fink and Walford, Int. Comp. Biology 46(6), (2006)

[66] Hanegan, J.L. and J.E. Heath, J. Exp. Biol. 53, (1970)

[67] Altizer, S. and C. Bradley, at http://www.uga.edu/monarchparasites/research/index.html#flight

[68] Alexander, David *Nature's Flyers: Birds, Insects and the Biomechanics of Flight,* (2002)

[69] Discover, April, (1997)

[70] Env. Entom. (33), (2004)

[71] Schuettler, Ranelli and Kelly, http://www.monarchlab.org/research/strom1poster.pdf

[72] Davis and Garland, in *The Monarch Butterfly: Biology and Conservation,* (2004)

[73] Calvert, J. Lepidopt. Soc. 53(1), (1999)

[74] http://www.texasento.net/MonarchMigration.htm

[75] Calvert and Wagner, in *North American Conference on the Monarch Butterfly* (1999)

[76] http://www.monarchlab.umn.edu/research/Rep/diapause.html

[77] Lynch and Martin, in *Biology and Conservation of the Monarch Butterfly* (1993)

[78] Anderson & Brower, Ecol. Entomol. 21(2), (1996)

[79] Troyer, Burks and Lee, J. if Insect Physiology, 42(7), (1996)

[80] Brower and Calvert, Evolution 39(4), (1985)

[81] Fink and Brower, Nature 291, (1981)

[82] http://www.monarchlab.umn.edu/research/Mwd/mwd.html

[83] Glendinning, Chemoecol. 1(3-4) (1990)

[84] Altizer, Oberhauser and Brower, Ecol. Entomol., 25 (2), (2000)

[85] Bradley and Altizer, Ecol. Letters 8, (2005)

[86] Leong, *et al*, Ecol. Entomol. 17(4), (1992)

[87] Lindsey *et al,* Ecol. Entomol. 34(5), (2009)

[88] Roode, J., Yates, A. and Altizer, S. PNAS 105, (2008)

Chapter 5 Notes

[1] Brower, L.P. in *North American Conference on the Monarch Butterfly* (1999)

[2] Brower, L.P. in *North American Conference on the Monarch Butterfly* (1999)

[3] Alonso-Mejia *et al*, Biol. Cons. 85(1) (1998)

[4] Brower, L.P. in *North American Conference on the Monarch Butterfly* (1999)

[5] Adriana Alatorre. Reforma. Mexico City: Feb 25, (2008)

[6]

http://earthobservatory.nasa.gov/Newsroom/NewImages/images.php3?img_id=17943,

http://www.nytimes.com/2008/03/07/science/earth/07butterfly.html

[7] Maza and Calvert (*Biology and Conservation of the Monarch Butterfly* (1993)

[8] http://www.ens-newswire.com/ens/feb2002/2002-02-14-03.asp

[9] Rendon-Salinas *et al*,

http://www.wwf.org.mx/wwfmex/descargas/rep_monitoreo_colonias_monarca_06-07.pdf

[10] Rashin, Small and Alcala, in *North American Conference on the Monarch Butterfly* (1999)

[11] Brower *et al* Cons. Biol. 16(2), (2002)

[12] Ramirez, Azcarate and Luna (Forestry Chron. 79(2), (2003).

[13] Ogarrio in *Biology and Conservation of the Monarch Butterfly* (1993)

[14] Alonsio-Mejia *et al* Eco. Appl. 7(3), (1997)

[15] Honey-roses, J. Land Degrad. Develop. 20, (2009) and online at: http://jhoney.googlepages.com/honey-roses2009.pdf

[16] Homero Aridjis, "Reforma," Nov. 6, (2005)

[17] Snook in *Biology and Conservation of the Monarch Butterfly* (1993)

[18] Davis, H., Journal of Insect Conservation p. 1-8 (2008)

[19] http://www.commondreams.org/headlines03/1111-06.htm

[20] Butler, Vickery and Norris , Science 315, (2007)

[21] Malcolm in *Biology and Conservation of the Monarch Butterfly* (1993)

[22] (Desneux, Decourtye and Delpeuch, Annu. Rev. Entomol. (52), (2007)

[23] Oberhauser, *et al*, J. Am. Mosq. Control Assoc. 25(1):83-93, (2009)

[24] Oberhauser, *et al*, Env. Entomol. (35)(6), (2006)

[25] Losey, *et al*, Nature 399, (1999)

[26] Biochimie 84. (2002)

[27] Stanley-Horn *et al* Proc. Natl. Acad. Sci. USA 98(22) (2001)

[28] Gathmann *et al*, Mol. Ecol. 15(9), (2006)

[29] Proc. Natl. Acad. Sci. USA 98, (2001)

[30] Trends in Genetics 18(5), (2002)

[31] Prasifka, *et al*, Env. Entomol. (36), (2007)

[32] http://www.monarchwatch.org/update/2006/0930.html#2

[33] Chip Taylor, Dplex-l list, citing acreage from
http://usda.mannlib.cornell.edu/usda/current/Acre/Acre-06-29-2007.pdf

[34] Kline, L.J. *et al*, Northeastern Naturalist 15(1) (2008)

[35] http://www.guardian.co.uk/environment/2009/apr/27/butterfly-decline-
conservation-endangered-species

[36] Warren, *et al*, Nature, 414, (2001)

[37] *Daily Telegraph,* Feb. 11, (2007)

[38] Marisco. Am. Midl. Nat. 154(2), (2005)

[39] McKenna, *et al*, J. Lepid. Soc. 55(2), (2001)

[40] http://tncweeds.ucdavis.edu/esadocs/documnts/vinc_sp.pdf

[41] Casagrande, R.A. and Dacey, J.E., Env. Entomol. (36), (2007)

[42] Mattila and Otis. Ent. Exp. et Appl. 107(3), (2003)

[43] Losey, Ent. Exp. et Appl. (2003)

[44] Calvert, J. Lepidopt. Soc. 53(1), (1999)

[45] http://tncinvasives.ucdavis.edu/

[46] Koch *et al*, Biol. Invasions 8(5), (2006)

[47] Koch *et al*, Environ. Entom. 34(2) 2005)

[48] Rudolph, D.C. *et al*, J. of the Lepidopterists Society, 60(3), (2006)